Biological Complexity and Integrative Pluralism

This fine collection of essays by a leading philosopher of science presents a defense of integrative pluralism as the best description for the complexity of scientific inquiry today. The tendency of some scientists to unify science by reducing all theories to a few fundamental laws of the most basic particles that populate our universe is ill-suited to the biological sciences, which study multicomponent, multilevel, evolved complex systems. This integrative pluralism is the best way to understand the different and complex processes – historical and interactive – that generate biological phenomena.

This book will be of interest to students and professionals in the philosophy of science.

Sandra D. Mitchell is Professor in the Department of History and Philosophy of Science at the University of Pittsburgh.

CAMBRIDGE STUDIES IN PHILOSOPHY AND BIOLOGY

General Editor
Michael Ruse   *Florida State University*

Advisory Board
Michael Donoghue   *Yale University*
Jean Gayon   *University of Paris*
Jonathan Hodge   *University of Leeds*
Jane Maienschein   *Arizona State University*
Jesús Mosterín   *Instituto de Filosofía (Spanish Research Council)*
Elliott Sober   *University of Wisconsin*

Published Titles

Alfred I. Tauber   *The Immune Self: Theory or Metaphor?*

Elliott Sober   *From a Biological Point of View*

Robert Brandon   *Concepts and Methods in Evolutionary Biology*

Peter Godfrey-Smith   *Complexity and the Function of Mind in Nature*

William A. Rottschaefer   *The Biology and Psychology of Moral Agency*

Sahotra Sarkar   *Genetics and Reductionism*

Jean Gayon   *Darwinism's Struggle for Survival*

Jane Maienschein and Michael Ruse (eds.)   *Biology and the Foundation of Ethics*

Jack Wilson   *Biological Individuality*

Richard Creath and Jane Maienschein (eds.)   *Biology and Epistemology*

Alexander Rosenberg   *Darwinism in Philosophy, Social Science and Policy*

Peter Beurton, Raphael Falk, and Hans-Jörg Rheinberger (eds.)   *The Concept of the Gene in Development and Evolution*

David Hull   *Science and Selection*

James G. Lennox   *Aristotle's Philosophy of Biology*

Marc Ereshefsky   *The Poverty of the Linnaean Hierarchy*

Kim Sterelny   *The Evolution of Agency and Other Essays*

William S. Cooper   *The Evolution of Reason*

Peter McLaughlin   *What Functions Explain*

Bryan G. Norton   *Searching for Sustainability*

Biological Complexity and Integrative Pluralism

SANDRA D. MITCHELL

University of Pittsburgh

CAMBRIDGE
UNIVERSITY PRESS

PUBLISHED BY THE PRESS SYNDICATE OF THE UNIVERSITY OF CAMBRIDGE
The Pitt Building, Trumpington Street, Cambridge, United Kingdom

CAMBRIDGE UNIVERSITY PRESS
The Edinburgh Building, Cambridge CB2 2RU, UK
40 West 20th Street, New York, NY 10011-4211, USA
477 Williamstown Road, Port Melbourne, VIC 3207, Australia
Ruiz de Alarcón 13, 28014 Madrid, Spain
Dock House, The Waterfront, Cape Town 8001, South Africa

http://www.cambridge.org

First published 2003

Printed in the United States of America

Typeface Times Roman 10.25/13 pt.   *System* LATEX 2_ε [TB]

A catalog record for this book is available from the British Library.

Library of Congress Cataloging in Publication Data

Mitchell, Sandra D., 1951–
Biological complexity and integrative pluralism / Sandra D. Mitchell.
   p.   cm. – (Cambridge studies in philosophy and biology)
Includes bibliographical references (p.).
ISBN 0-521-81753-6 (hb) – ISBN 0-521-52079-7 (pb)
1. Biological systems.   I. Title.   II. Series.
QH331 .M49   2003
570–dc21   2002038843

ISBN 0 521 81753 6 hardback
ISBN 0 521 52079 7 paperback

For Joel Murray Smith, my inspiration

Contents

Contents

List of Tables and Figures

TABLES

FIGURES

List of Tables and Figures

Preface and Acknowledgments

This collection of essays defends integrative pluralism as the best description of the relationship of scientific theories, models, and explanations of complex biological phenomena. Complexity is endemic in biology, but it is constituted by various features of multicomponent, multilevel, evolved systems. The types of scientific representations and the very methods we use to study biological systems must reflect both that complexity and variety. Developing models of single causal components, such as the effects of genetic variation, or of single-level interactions, such as the operation of selection on individuals, give valuable, if partial, accounts. These explanations need to be integrated in order to understand what historical, proximal, and interactive processes generate the array of biological phenomena we observe.

Clearly, the way the world is dictates what we can say about it. The way our representations are structured also plays a significant role in the scientific accounts we develop. Theories and models are idealized, partial descriptions, couched in the conceptual frameworks of the day, framed in a language that carries meanings from the broader social context. The suggestion that our current best theories of the nature of nature exactly capture the world in all its details is hubris. The idealized and partial character of our representations suggest that there will never be a single account that can do all the work of describing and explaining complex phenomena. Different degrees of abstraction, attention to different components of a system, are appropriate to our varying pragmatic goals and conceptual and computational abilities. In short, both the ontology and the representation of complex systems recommend adopting a stance of integrative pluralism.

I have developed the ideas and arguments in this book over a period of fifteen years. There are many people who have had important influences on the way I think about these issues. Naturally, my early teachers in philosophy of science – Jim Bogen, Imre Lakatos, and Peter Machamer – get some of the blame. In addition, Nancy Cartwright has been both friend and mentor. The philosophical work of John Dupré, Elliott Sober, and Bill Wimsatt have also been stimulating.

My deepest thanks must go to Rob Page. We met in the 1980s at the Ohio State University where he was in the Department of Entomology. Rob has enormous enthusiasm for his science, astuteness in his research, and a desire to get it right. He has opened the door for me to get an inside look at biology at its best, and I have learned a great deal from him. Indeed, Rob is coauthor of Chapter 3, section 3.1, "The Evolution of Division of Labor," and also of "Idiosyncratic Paradigms and the Revival of the Superorganism," one of the papers on which Chapter 2, section 2.1, is based. The other scientists whose ideas have been important to my approach are Steve Gould and Stuart Kauffman.

I am grateful for the support and challenge provided by my colleagues and students at the University of Pittsburgh and previously at the University of California. In addition, the years I spent at the Center for Interdisciplinary Research in Bielefeld, Germany, and at the Wissenschaftskolleg in Berlin provided intellectual opportunities for which I am grateful. The Santa Fe Institute has also been a place for new ideas and new collaborations.

I thank Megan Delehanty, Dennis Pozega, and Melissa Wurster for help in the production of the book, and Michael Ruse for suggesting it in the first place.

I have dedicated this book to Joel Smith, my husband, my friend, and the best critic and supporter one could wish for.

ACKNOWLEDGMENTS

The research for the ideas in this book was funded by the National Science Foundation, Program in Science and Technology Studies, the Santa Fe Institute, the Center for Interdisciplinary Studies in Bielefeld, the Wissenschaftskolleg in Berlin, and grants from the University of California.

The introductions to each of the chapters have not been previously published. Also new for this volume is the first section of Chapter 1 and Chapter 6, section 6.1, "Critics of Unity of Science." There are two sections that are the result of merging two previously published articles. Chapter 2, section 2.1, "Compositional Complexity and the Superorganism Metaphor," is from Sandra Mitchell, "The Superorganism Metaphor: Then and Now," in S. Maasen, E. Mendelsohn, and P. Weingart, eds., *Biology as Society, Society as Biology: Metaphors*, Yearbook in the Sociology of Science

(Dordrecht: Kluwer Academic Publishers, 1995), 231–248, and "Idiosyncratic Paradigms and the Revival of the Superorganism," coauthored by Sandra Mitchell and Robert E. Page, Jr., *Report NR. 26/92 of the Research Group on Biological Foundations of Human Culture*, Bielefeld, Germany, 1992. Chapter 4, section 4.3, "On Biological Functions," is from "Dispositions or Etiologies: A Comment on Bigelow and Pargetter," *Journal of Philosophy* 40, no. 5 (May 1993): 249–259, and "Function, Fitness and Disposition," *Biology and Philosophy* 10 (1995): 39–54.

The other sections are reprinted as follows: Chapter 3, section 3.1, from R. E., Page, Jr., and S. D. Mitchell, "Self Organization and the Evolution of Division of Labor," *Apidologie* 29 (1998): 101–120; Chapter 4, section 4.1, from "Competing Units of Selection? A Case of Symbiosis," *Philosophy of Science* 54 (1987): 351–367; section 4.2, from "Units of Behavior in Evolutionary Explanations," in Marc Bekoff and Dale Jamieson, eds., *Interpretation and Explanation in the Study of Animal Behavior* (Boulder: Westview Press, 1990), 63–83; Chapter 5, section 5.1, from "Pragmatic Laws," *Philosophy of Science* 64 (1997): S468–S479; section 5.2, from "Dimensions of Scientific Law," *Philosophy of Science* 67 (2002): 242–265; section 5.3, from "Contingent Generalizations: Lessons from Biology," in R. Mayntz, ed., *Akteure, Mechanismen, Modelle: Zur Theoriefähigkeit makrosozialer Analysen* (Frankfurt: Campus, 2000), 179–195; section 5.4, from "Ceteris Paribus – An Inadequate Representation for Biological Contingency," *Erkenntnis* 57, no. 3 (2002): 329–350; Chapter 6, section 6.2, from "On Pluralism and Competition in Evolutionary Explanations," *American Zoologist* 32(1992): 135–144; and section 6.3, from "Integrative Pluralism," *Biology and Philosophy* 17 (2002): 55–70.

1

Introduction

In a variety of scientific disciplines, political discussions, and social policy forums, there is increasing interest in diversity as an inherent good. Stifling the voices or participation of underrepresented ethnic or economic groups is tantamount to giving up democratic ideals. While there have been trends toward "big science," collapsing scientific research into one or a small number of tracks, diversification of projects funded is associated with potential for creative breakthroughs. Indeed, the National Science Foundation's mission statement acknowledges, "The needs and opportunities of the science and engineering enterprise come in all shapes and sizes. The challenge to NSF is to meet these needs and pursue these opportunities in ways that are appropriate in each case" (National Science Foundation 1995). Since the establishment of the Convention on Biological Diversity, the variety and variability of life itself has been deemed of "intrinsic value." Yet the attitude of deeming diversity of any kind a prima facie good has prompted justified concern about sliding down a slippery slope to complete relativism with an accompanying loss of critical standards in all these domains.

In science, the debate is reflected in a dispute about the unity or disunity of science. In this context, reductionists hold a set of beliefs and methodologies aiming to reduce the diversity of explanations to a small number of theories or laws at a privileged level of discourse, thereby globally unifying science. This standpoint has been resisted by many who investigate higher-level phenomena and represent the knowledge gleaned of psychological properties or biological properties in ways that appear to be devalued by reductionism. They support some type of pluralism and disunity. Philosophers of science have attempted to help settle the debate by clarifying the grounding of both sides. Unity is apparently entailed by strong critical standards based on metaphysical and methodological monism (see Chapter 6, section 6.1). And yet, the reduction of psychology to neurobiology or biology to chemistry and

1

chemistry to physics has never been realized in practice. This failure of philosophical analyses that support unity through reductionism and monism to match the practice of science has led some to consider alternatives to unity. However, disunity can seem to rest on forms of pluralism that verge on an uncritical, "anything goes" leniency (see Galison and Stump 1996 for a range of pluralistic views). Even if one rejects the extreme pictures of a rigorous, reduced, and unified science or a relativistic, pluralistic, and disunified science, there are many places to occupy in the middle ground. Which pluralistic picture of scientific practice should be endorsed, and what are the critical standards retained? The following chapters investigate that territory and defend a framework for a critical pluralism. The main argument is that the complexity of the subjects studied by the various sciences and the limitations of our representations of acquired knowledge jointly entail an integrative, pluralistic model of science. It is a model that allows but does not demand reduction. It is a model that recognizes that pluralist ontologies and methodologies are required to give an accurate picture of science as currently understood and practiced.

The "fact" of pluralism in science is no surprise. On scanning contemporary journals, books, and conference topics in some sciences, one is struck by the multiplicity of models, theoretical approaches, and explanations for the phenomena of interest. For example, there is a host of alternative models and explanations for division of labor in social insects. Division of labor refers to an intricate set of behaviors that may involve thousands of individual ants, bees, or termites engaged in a variety of tasks, including cleaning, raising the brood, tending to the queen, building a comb or nest, guarding, and foraging. It has long been known that there is a temporal component to the tasks; younger workers perform in-nest tasks, older individuals do outside jobs. Some workers specialize and others are generalists, and the colony as a whole responds to changing needs by reallocating workers to the jobs that address those needs. Accounts of division of labor have included group, individual, and "selfish gene" adaptation explanations that appeal, respectively, to claims of ergonomic or other forms of optimization (Oster and Wilson 1978; Wilson 1971; Winston 1987); proximate mechanisms of juvenile hormone regulation and the development of specialized morphological features (Robinson et al. 1989); and variant models of emergent self-organization (Deneubourg et al. 1987; Page and Mitchell 1991; Page and Robinson 1991; Robinson and Page 1989a; Seeley 1989, 1995; Tofts and Franks 1992).

If science is representing and explaining the structure of the *one* world, it is reasonable to ask why there is such a diversity of representations and explanations in some domains. One response is that this simply reflects the immaturity of the science (Kuhn 1962). Yet history shows us that many sciences never exhibit a diminution in the multiplicity of theories, models, and explanations they generate. This "fact" of pluralism, on the face of it, seems to be correlated not with the maturity of the discipline but with the complexity of the subject matter. Compared with theories of social insect biology, there appear to be fewer alternative models of chemical bonding or particle dynamics, while studies of complex human social behavior display an even larger range of alternative views. Rather than explaining away the current pluralism as symptomatic of immaturity, we should invert the order of allegiance. The diversity of views is not an embarrassment or sign of failure, but rather the product of scientists doing what they must do to produce effective science. Pluralism can simply reflect complexity. But what type of pluralism? What type of complexity?

Pluralism in scientific theory and practice has certainly been studied and advocated before. As a result of historical investigations of scientific methods, Feyerabend was led to defend epistemological anarchism (1975, 1981). He endorsed any method as being as good as every other in generating acceptable scientific results. Like Feyerabend's, the most recent defenses of pluralism in science have launched their accounts from an epistemological perspective. Feminist stance theory, for example, grounds pluralism in a perspectivalism based on individual and group social experience (Harding 1986; Longino 1990). Other disunity supporters argue from the partial character of descriptions or diverging areas of interest of the researchers (Cartwright 1980, 1982, 1989, 1994; Dupré 1983, 1996). In contrast, a few have linked pluralism with ontological views about natural kinds (Dupré 1983, 1993; Hacking 1996). While my approach draws on many of the insights of these philosophers, it differs from them in

- grounding pluralism jointly on metaphysical and epistemological arguments, and
- appealing to complexity as a critical tool for understanding the nature and limits of diversity in scientific methods and representations.

The conceptual framework developed in this collection of essays has three practical results: (1) it should allow one to better sort between spurious and real disputes in science; (2) it should provide a taxonomy of typical kinds of conflict, emphasizing a distinction between representational and substantive conflict; and (3) it should allow a better understanding of different kinds of compatibility and complementarity between alternate theories.

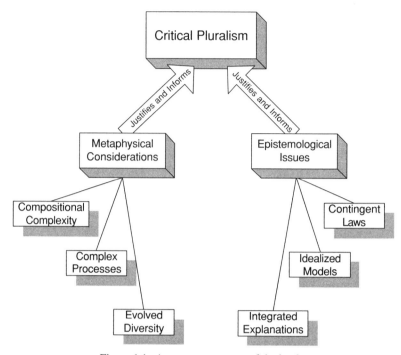

Figure 1.1. Argument structure of the book.

Figure 1.1 shows the basic structure of the argument of the book.

PART I: COMPLEXITY

The first part of the book explores the left side of this picture, that is, the nature of complexity in the world. Current definitions of complexity number somewhere between 30 and 45 (at least according to Horgan's report of Yorke's list; see Horgan 1996: 197, n. 11; see also *Science* 284 (2 April 1999). The multiplicity of definitions of "complexity" reflects not confusion on the part of scientists but the actual variety of ways that systems are complex. In this chapter I outline a taxonomy of complexity for biology. In Part I of the book I explore some characteristics of these different species of complexity, with special attention to the example of social insect colonies. Complexity can be categorized as follows:

- *Constitutive complexity:* Organisms display complexity of *structure*, the whole being formed of numerous parts in nonrandom organization (Simon 1981; Wimsatt 1986).

4

- *Dynamic complexity:* Organisms are complex in the *processes* by which they develop from single-celled origins to multicellular adults (Goodwin 1994; Goodwin and Saunders 1992; Raff 1996) and by which they evolve from single-celled ancestors to multicellular descendants (Bonner 1988; Buss 1987).
- *Evolved complexity:* The *domain* of alternative evolutionary solutions to adaptive problems defines a third form of complexity. This consists of the wide diversity of forms of life that have evolved despite facing similar adaptive challenges.

I discuss each of these categories in brief detail here and develop these ideas further in the chapters that follow.

A. Compositional Complexity

Minimally, complex systems can be distinguished from simple objects by having multiple parts that stand in nonsimple relations. That is, there is structure or order in the way in which the whole is composed of the parts. Individual cells constituting a multicellular organism differentiate into cell types and growth fields (Raff 1996). The collections of species that form an ecosystem occupy different niches and perform different functions for sustaining ecosystem integrity (Krebs 1994). Individual insects that make up a colony behaviorally differentiate into task specialists and age castes (Wilson 1971; Winston 1987). Indeed, invoking the analogy between individual organism and social insect colony has a long history marked by the use of the superorganism metaphor. In Chapter 2, I argue that invoking this metaphor for explanatory purposes may carry with it conceptual baggage that obscures the variety of compositional structures that describe social insect organization. Indeed, a similar variability in composition rules applies to individual organisms (Buss 1987).

Ecosystems, too, constitute compositionally complex systems, consisting of various nonliving, abiotic, and living, biotic components. The abiotic components of an ecosystem include various physical and chemical factors, while the biotic components are the array of species and organisms present. There are clearly different ways to compose an ecosystem, with different species occupying similar niches or a different total number of components, while maintaining the functioning of nutrient and energy cycling through the system. The variability of compositional complexity generates problems for representing such systems in scientific models and explanations (see also Wimsatt 2000).

B. Complex Dynamics

Complexity has recently become closely associated with nonlinear mathematical functions representing temporal and spatial processes. This type of process complexity is linked with a number of dynamical properties, including extreme sensitivity to initial conditions, self-organizing and recursive patterning (e.g., thermal convection patterns), and negative and positive feedback regimes (amplification and damping) (Nicolis and Prigogine 1989). The striking discoveries of the generality of the models of complex dynamical processes found in basic chemical and physical systems has led to their increasing application to biological systems. For example, self-organization processes are ones in which higher-level order emerges from the simple interactions of component parts in the absence of a preprogrammed blueprint. The bifurcation of ant trails to a food source (Nicolis and Deneubourg 1999) and the emergence of division of labor (Page and Erber 2002) are examples. Holland's theory of complex adapted systems and Kauffman's dynamical models are attempts to describe generic features of systems that engage in such processes (Holland 1995; Kauffman 1993, 1995).

Simulations and empirical investigations into the behavior of social insects investigate these dynamics at a more concrete level (Deneubourg et al. 1987; Page and Mitchell 1991; Tofts and Franks 1992). In Chapter 3, I present models of self-organization for honeybee division of labor developed by Robert E. Page, Jr., and me. Self-organization refers to any set of processes in which order emerges from the interactions of the components of system without direction from external factors and without a plan of the order embedded in any individual component. The mechanisms of self-organization might be analyzed in terms of negative and positive feedback regimens. In Chapter 6, section 6.3, I compare the Page and Mitchell models with other self-organization models of division of labor in social insects and argue that which models are applicable depends on specific evolved features of the system. It is here that I defend the view that a plurality of models may be required to explain what is taken to be the "same" feature of the world instantiated in different individual systems.

Other complex dynamics of bifurcation and amplification are common in complex biological systems. In particular, ecosystems, with their multiple causal components, display discontinuous change that can be modeled by nonlinear equations (Holling 1986). As disturbing influences on an ecosystem increase, they can induce a major change in the organization and functioning of the system in unpredictable ways. A small increase in the disturbance, under certain conditions, can generate catastrophic effects, while at other times generate only correspondingly small changes in the system. An example is in coral reef degradation. There appear to be two stable equilibria states, one dominated by coral, the other dominated by macroalgae (Hatcher et al. 1989). The switch between these two states, a bifurcation, has been identified as a phase change (Done 1992) and has multiple causes operating at different time scales (e.g., storms and toxic pollutants, as well as long-term consequences of overfishing). The dynamic complexity is displayed by the fact that disturbed areas of coral ". . . can recolonize without passing through a macroalgal phase, or they may enter a macroalgal phase which they retain indefinitely" (Done 1992: 124).

C. Evolved Diversity of Populations

The third sense of complexity found in biology is exhibited by the diversity of organisms resulting from their historically contingent evolution. Given the irreversible nature of the processes of evolution, the randomness with which mutations arise relative to those processes, and the modularity by which complex organisms are built from simpler ones, there exists in nature a multitude of ways to "solve" the problems of survival and reproduction. Earlier "solutions" have downstream effects on the types of evolutionary change accessible in the future.

Ecology, again, provides cases. The classic example is of marsupial species populating Australia while placental species of similar ecological role, such as the kangaroo and the antelope, are found elsewhere. Why are marsupials so concentrated in range? The best hypothesis is that marsupials constitute a stable body plan that originated in Australia and came to concentrate on that continent as a result of land separation that preceded their evolution. The convergent evolution of similar placental and marsupial species marks different, equally good, solutions to similar selective environments.

In Chapter 3 I detail the structure of evolutionary explanations that account for this type of diversity of adaptation. There is a concomitant diversity in the scientific accounts of adaptation. The alternative representational frameworks we have in which to record the evolutionary history of life on the planet, as well as the different explanatory questions that are raised in the scientific exploration of that evolved diversity, generate some of the pluralism we observe in contemporary biology. The discussion of genic versus multilevel selection models concerns causal complexity, while the discussion of biological function and its multiple meanings investigates methodological pluralism.

PART II. PLURALISM

The second strand of the argument for critical pluralism focuses on the episte-mological and cognitive components of scientific knowledge. The aim for the unity of science draws much intuitive support from a picture of the world as ordered by laws. If we were to find the presumably few, basic laws that govern the events that unfold in our world, we would be in a position to effi-ciently explain, predict, and intervene successfully in that world; so the unity of science by reduction claims. The traditional philosophical accounts expli-cate knowledge of laws as knowledge of contingent, universal, exceptionless truths. Thus, one would reason, even if the world were complex, the under-lying organizing causal structures are not. If science actually discovered the kind of laws that philosophers have so carefully considered, then we would be able to do all the scientific chores that are described above. But such laws are few and far between. In Chapter 5, I develop an alternative to the traditional philosophical picture of laws, which I identify as a pragmatic conception of laws. This account has the following virtues. It both is a more accurate ac-count of actual scientific knowledge claims and provides a richer conceptual framework in which to locate and compare the variety of knowledge claims that constitute scientific knowledge. My argument here rests on both the meta-physical complexity detailed in the first part of the book, the idealization and partiality that characterize our representations of knowledge, as well as the diverse interests that are served by that knowledge. My pragmatic conception of laws does not rule out universal, exceptionless regularities but broadens the concept of "law" to one that is both more accurately descriptive of what scientists use for prediction and explanation and is compatible with critical pluralism.

In Chapter 6 the question of pluralism versus unity in science is considered directly. Clarification of the places to stand in the unity/disunity terrain is pro-vided. Here I consider an existing model of pluralism in biology, Sherman's (1988) levels of analysis. His approach is a development of Ernst Mayr's (1961, 1982) distinction between "how" and "why" questions and Nikolaas Tinbergen's (1963) four questions for ethology. A levels of analysis model of pluralism acknowledges that different types of questions require different answers and hence might not be in conflict. While there are valuable insights in this approach, the levels of analysis framework fails to adequately rep-resent the relations between alternative explanations. It misconstrues where conflict does and should occur and where alternatives are correctly judged to be compatible. In the extreme case, it can lead to a form of isolationism that can impede answering questions within any single level. The mistake lies not in recognizing a diversity of questions – indeed, scientists do pose a variety of questions to the subjects they study (see Van Fraassen 1980) – but rather in the assumptions made about the epistemological structure of the answers.

A more promising model of pluralism can be forged from understanding that causal models are abstractions that will always remain idealizations. By making simplifying assumptions regarding the noninterference of other potential causes, causal models describe only what would be expected in idealized circumstances (Levins 1968). This conception of theories helps to explain how it can be that models of different causal factors qua models do not directly conflict. In addition, I argue that even if the models may be jointly consistent, in the application of models to the explanation of a concrete case, conflict can arise (see also Chapter 2). In actual cases, multiple causes are likely to be present and interact, and other local elements may also contribute to a specific causal history. Thus in explanation, models of variant possible contributing factors must be integrated to yield the correct description of the actual constellation of causes and conditions that brought about the event to be explained.

Chapter 6 also investigates the philosophical defense of disunity developed by John Dupré (1993, 1996). He argues that metaphysical materialism is, by itself, insufficient to support unity of science by reduction. He suggests that the addition of an assumption of causal completeness at the physical level is required to draw the conclusion that all sciences are ultimately reducible. Yet he argues that we should reverse the inference and take the failure of successful reduction to lead us to reject causal completeness. While I agree with his rejection of reductionism, I argue that his analysis fails to sufficiently acknowledge the nature of scientific representations. In contrast, I suggest that while a well-grounded critical pluralism is materialistic, it is structured by both metaphysical complexity and features of our representations of that complexity.

The investigations in the chapters that make up this book allow the development of a constrained, critical pluralism of scientific theories and explanations. The debate between global unity of science through reduction to a single representational framework or ontological level versus a disunity of incommensurable alternatives is then recast to allow a more refined assessment of where unity can and should be forged and where diversity can and should be championed. In conclusion, critical pluralism as a perspective about the shape of scientific practice is defended. Should the current diversity of laws, models, and explanations in science be viewed as grist for the unifier's mill, raw material to be worked into a single, unified theory of everything? Or should some form of pluralism be not only expected but required for effective scientific practice? This book answers these questions through an analysis of the complexity found in nature and the structure of the representations that scientists devise in order to make sense of that complexity. In the end, the question of the unity or disunity of science can be more rigorously posed and, therefore, more adequately answered. Unification takes place by means of the integration of different theories and models that address partial causes that contribute to the generation of biological phenomena. It is not always to be found in a decreasing number of explanatory theories, but in many cases, quite the contrary. As we decompose multicomponent, multilevel, evolved complex systems into their constituent parts and processes and study these in more detail, more, rather then fewer, models will be the consequence.

I

Complexity

2

Constitutive Complexity

The complexities of organization or structure have implications for the scientific theories and models that we devise to represent them. Which arrangements of parts are stable through time? Which are capable of evolving over time? And which will be selected for in light of their consequences on the fitness of the parts and the wholes made up of those parts? One of the lessons of the history of life on this planet is that complexity, at least if defined as the number of cell types in a organism (Bonner 1988, 1998), has increased. Multicellular organisms evolved from single-celled organisms, more differentiated organisms evolved from those with fewer specialized cell types. Some changes of this type take place in the span of a life-cycle, for example, in the slime mold *Dictyostelium discoideum*, which begins as a collection of single amoeba and develops into a multicellular, differentiated individual with stalk and fruiting bodies.

A parallel might be drawn to the development of social groups, as in social insects. Here individual ants, bees, and wasps join together to form a colony where different individuals perform different functions, such as food acquisition and reproduction.

In this chapter I argue that the parallel between the evolution and development of multicellular organisms and that of social organization of insect colonies can be misleading if the diversity found in both types of organization is overlooked.[1]

---

[1] This chapter is the product of merging two previously published papers: S. D. Mitchell, "The Superorganism Metaphor: Then and Now," in S. Maasen, E. Mendelsohn, and P. Weingart, eds., *Biology as Society, Society as Biology: Metaphors. Yearbook in the Sociology of Science* (Dordrecht: Kluwer Academic Publishers, 1995), 231–248, and S. D. Mitchell and Robert E. Page, Jr., "Idiosyncratic Paradigms and the Revival of the Superorganism," report NR. 26/92 of the Research Group on Biological Foundations of Human Culture, Bielefeld, Germany, 1992.

2.1. COMPOSITIONAL COMPLEXITY AND THE
SUPERORGANISM METAPHOR

Our choice of models, and to some extent our choice of words to describe them, is important because it affects how we think about the world. . . . [O]ur choice of model decides what phenomena we regard as readily explicable, and which need further investigation. (Maynard Smith 1987: 120)

Introduction

The complex behavior of termite, wasp, ant, and bee societies, with their division of labor and coordinated activity, has long been a subject of wonder and awe. Yet the cooperation among individual workers and their functional sterility, combined with helping the queen to produce offspring, were seen even by Darwin (1859) as problematic for a theory built around the competitive struggle for existence between individual organisms. Darwin resolved the problem, to his own satisfaction at least, by invoking a higher level of selection, that of the colony. Recently, there has been an attempt to revive colony-level selection as a special form of group selection, by reviving an old metaphor, that of the superorganism (Wheeler 1911), and to sophisticate it as a "theory of superorganisms" for social insects (Lumsden 1982; Seeley 1989; Wilson 1985b; Wilson and Sober 1989). The original application of a superorganism theory to explain social insects was posed as an addition to Darwin's theory in which " . . . the struggle for existence is not more than half the truth. . . . To us it is clear that an equally pervasive and fundamental innate peculiarity of organisms is their tendency to cooperation or 'mutual aid'" (Wheeler 1923: 3). In this essay I examine the renewed advocacy for the superorganism theory and argue against its revival. Unlike the original model of Wheeler and E. O. Wilson's extension of that model, the theory as reformulated by D. S. Wilson and Sober (1989) and applied by Seeley (1989) rests on what we believe is an idiosyncratic paradigm that does not adequately represent existing entities. We argue that it is deficient in two ways. First, the commitment to a particular form of functional organization (ideal Weismannism) obscures the variety of ways in which social insect societies are in fact organized. Second, the selectionist criteria for identifying functional organization at the colony level obscure the role of other potential sources of organization.

There are two contemporary approaches to the superorganism theory. While both approaches coopt the name "superorganism," they make use of conflicting accounts of the defining characteristics of a superorganism. The conflation of these accounts as a result of the common terminology tends

14

to obscure very important differences. We argue that while there are merits in each, it is a mistake to reintroduce the superorganism concept to legitimate hierarchical selection theory or to emphasize the role developmental processes play in generating the observed features of social insects. Both the original and revived superorganism theories invoke a narrow definition of an organism that is based on the Weismannian organizational paradigm of complete separation of germ plasm and somatic cells. This Weismannian ideal has been challenged as unsuitable for describing individual organisms in most taxa (Buss 1987), and we suggest that its superorganismic counterpart is equally deficient in describing most social insects. We believe that a more general theory in terms of complex systems allows both goals to be met without invoking a concept that carries its own history, inappropriate to the task at hand, and that lends itself to misuse. Of course, whether one uses the term "superorganism" or not is not at issue. One can clearly retain the term, but only by decoupling it from its historical usage and its Weismannian connotation, that is, by redefining it. However, we believe that the development of a general biological model in terms of hierarchical complex systems is more promising, since such modeling allows easier connection to even more general theoretical frameworks of complexity and order that are finding application in physics and chemistry as well as biology (Kauffman 1984; Nicolis and Prigogine 1989). If we construct biological theory on the basis of anthropocentric prejudices using idiosyncratic paradigms, we may preclude investigation and explanation of features shared by all complex systems, biological or otherwise.

The Superorganism

Superorganism metaphors and theories take the individual organism as a model of functional integration or cooperation of parts and extend that model to describe and explore social groups of individuals. So, just as individual cells in the body cooperate in the development, maintenance, and reproduction of the organism, so, too, it is suggested by the metaphor, do individuals cooperate in the development, maintenance, and reproduction of a colony or a society. Clearly, an organism such as a human being, with 46 chromosomes, approximately 210 different cell types, and more than 30,000 genes, strikes some resemblance to a social organization, like a honeybee colony with its four worker castes and tens of thousands of individual workers engaged in a variety of coordinated tasks. The ways in which complex biological organizations are structured is the proximate result of developmental processes and the historical result of evolution. Recently, there has been renewed interest in drawing the analogy between organisms and social groups by appeal to the superorganism metaphor. In what follows we critically assess the benefits as well as the costs of employing this choice of words in describing social insect societies.

Metaphorical transfers between organisms, social insects, and human societies has had a long history. Yet each historical and scientific context has stamped its own character on the use of such language. In particular, we discuss the recent arguments for the "revival" of the superorganism metaphor for the study of social insects. After decades when mention of superorganism was anathema in social insect studies (see Wilson 1968, 1971), in the 1980s there were multiple pleas to "revive" the superorganism. These include E. O. Wilson (1985a, 1985b), who claimed that although no one used superorganism language through the reductionistic and empirical trends in entomology from the '50s until the '80s, even then it had a significant, albeit "semi-conscious," role. In addition, articles supporting the revival include Charles Lumsden (1982), Thomas Seeley (1989), and David Sloan Wilson and Elliott Sober (1989).

The new interest in using the superorganism metaphor raises a number of questions. How does this choice of models and choice of words affect the way one studies social insects? In effect, what does a superorganismic revival for theories of social insects amount to? The answer to this analytic question is further complicated by two factors: One is that the words chosen are explicitly metaphorical and hence require the explication of the transfer of conceptual content between the primary and secondary contexts of application. One might call this the horizontal dimension of transfer. The second factor is that the plea is to revive a framework that had originally been voiced in a very different historical and scientific context. To see what the theoretical and empirical content of the so-called revived superorganism model is now, one must also investigate the vertical transfer of content from the antecedent application of the metaphor to its contemporary instantiation. The story of the revival of the superorganism metaphor thus concerns the transfer of language and content both from one scientific context (theories of the organism) to a second scientific context (theories of social groups) and from one historical period (early twentieth century) to another (late twentieth century).

Critical Theory of Metaphors

Earlier in the twentieth century, philosophers of science had tended to view metaphors and models as merely heuristic, nonessential, and hence dispensable for science. While perhaps relevant to the context of discovery, like hallucinations and other bogeymen of the peculiarly human psyche, it was thought that metaphorical language did nothing for explanation, justification, or the rational development of scientific theories. According to Bono, the "standard" view, developed in the scientific revolution in the seventeenth century, is that metaphors "introduce inappropriate, not literal meanings into science, contaminating . . . precise and stable meanings." That metaphorical language "compromises scientific inquiry and is to be avoided" (Bono 1990: 62). But this picture of science is being replaced by one in which metaphors are seen as ubiquitous, important, and powerful in scientific practice and hence demand critical analysis. "We need a critical theory of metaphor in science in order to expose the metaphors by which we learn to view the world scientifically, not because these metaphors are necessarily 'wrong'; but because they are so powerful" (Stepan 1986: 277). Philosophers have been engaged in the development of such a critical theory. The literature on this subject is vast, but we describe one trend within it that motivates the discussion of the organism metaphor that follows.

Metaphors were once understood to make explicit, definite claims of similarity or analogy between the primary reference or context and the secondary one. On this, the "comparison" view, communication with metaphorical language was possible only when all parties agreed with the specific similarity claims. With an admission of the less than precise and explicit nature of meaning necessary for communication, even for scientific discourse, this picture of metaphors has been replaced. Now, many philosophers defend a view of metaphor as more "open-ended" and "interactive" (Black 1979; Boyd 1979; Kuhn 1979).

Even with the loosening up of the criteria for metaphorical understanding, there are differences within this modern camp. Black, for example, claims that metaphors are useful only in pretheoretical stages of science, in pedagogical interactions, or in nontechnical popularizations. Boyd and Kuhn, however, find a place for metaphors in the construction and development of theory in mature science. Boyd refers to these as "theory constitutive" metaphors. Rather than require explicit similarity before a metaphor can be properly applied, Boyd claims that inductive open-endedness actually invites scientists to explore possible similarities. Metaphors are attempts to accommodate our language to not yet discovered causal features of the world. Kuhn, though differing from Boyd on the realistic or constructivist interpretation of such endeavors, agrees with this basic approach. Both endorse the view that, in Kuhn's words, "metaphor plays an essential role in establishing links between scientific language and the world . . . " (1979: 415–16). It follows that metaphors should not be left unjustified or "semi-conscious."

Rather, they, like any other theoretical postulate, should be accepted critically. Boyd says:

> One should employ a metaphor in science only when there is good evidence that an important similarity or analogy exists between its primary and secondary subjects. One should seek to discover more about the relevant similarities or analogies, always considering the possibility that there are no important similarities or analogies, or alternatively, that there are quite distinct similarities for which distinct terminology should be introduced. One should try to discover what the "essential" features of the similarities or analogies are, and one should try to assimilate one's account of them to other theoretical work in the same subject area – that is, one should *attempt* to explicate the metaphor. (1979: 406)

In addition to the horizontal comparisons between the two domains of application of a metaphor, a critical analysis of a specific metaphor requires the comprehension of the sources and implications of the use of the metaphor both within and outside the confines of scientific discourse. Metaphors, it is clear, are both enabling and constraining. While they structure our perceptions, allowing us to "see" causal structures in new domains, via the similarity to such structures in known domains, they at the same time proscribe certain observations, blind us to certain descriptions and awareness. While they serve as a program for research, they also run the danger of being confused for reality. And finally, while performing all these onerous tasks in scientific theorizing and practice, metaphors at the same time reflect the social structure, scientific organization, and cultural milieu in which they are invoked (see Mitman 1995). A critical analysis, ideally, should expose all of these aspects of a scientific metaphor. In this study, we undertake only part of this larger task.

In general, scientific theories make two sorts of ontological commitments in explaining phenomenological experience: one concerning which entities exist, and the other concerning which forces act upon those entities. The revival of the superorganism metaphor attempts to expand the ontology of Darwinian theory to include superorganisms as real, explanatory entities. We believe there are two different sources for the appeal to the reality of this ontological level, each resting on a different set of biological processes (see Table 2.1). E. O. Wilson (1971, 1985a, 1985b) and Lumsden (1982) describe superorganisms as entities subject to sociogenesis, the analog of morphogenesis, while Seeley (1989) and Wilson and Sober (1989) invoke superorganisms as entities subject to the forces of natural selection operating at a group level. Individual organisms are paradigmatic examples of both development and selection. They display ontogenetic processes of meiosis, mitosis, and cell differentiation and specialization – and hence a collection

Table 2.1. *Horizontal Transfer*

	Organism	Superorganism
Entities	Cells	Individuals
	Organism	Superorganisms
Organization	Weismann's preformationist development	Complete convergence of individual interests
Processes	Ontogenesis	Sociogenesis
	Selection on individual only	Selection on colony only

of individuals may be seen as a superorganism if it is subject to a similar ontogeny. As E. O. Wilson (1985a: 1492) says, "[t]he workers of advanced insect societies are not unlike cells that emigrate to new positions, transform into new types, and aggregate to form tissues and organs."

Individual organisms are also the paradigmatic subject of natural selection and the locus of adaptations – hence a collection of individuals may be seen as a superorganism if it is similarly subject to natural selection and its traits are adaptations at that level. Seeley thus claims,

> It seems correct to classify a group of organisms as a superorganism when the organisms form a cooperative unit to propagate their genes, just as we classify a group of cells as an organism when the cells form a cooperative unit to propagate their genes. (Seeley 1989: 548)

While the first approach emphasizes the developmental processes affecting the organism, the latter is concerned exclusively with how natural selection operates on organism-like entities. In this chapter, we consider both of these types of superorganism metaphor and argue that in each case, adopting this framework does more to obscure the biological phenomenon than to illuminate it. Before exploring these horizontal transfers, we first consider the progenitor metaphor developed by William Morton Wheeler in the early part of this century.

Vertical Transfer. The historical dimension of the investigation of the superorganism metaphor is required by the explicit desire on the part of contemporary scientists to "revive" the superorganism of W. M. Wheeler of the 1910s and 1920s rather than to introduce a newly coined, or newly framed, metaphor born of the contemporary scene. We suggest that there are significant differences between these periods that, we believe, are sufficient to cast doubt on the desirability of reviving the metaphor.

One way to compare the two scientific contexts is to see what elements make up the respective contrast classes for the superorganism defenders in each. By so doing, we do not intend to thereby promote the drawing of sharp dichotomies between naturalist and experimentalist, physiological and evolutionary questions (see Allen 1978) or vitalism and Darwinism (see Ghiselin 1974), but rather to also make apparent the range of views that allow for a middle ground. American biologists in particular tended to explicitly endorse the dissolution of dichotomies and the expansion of possibilities. For example, Charles Otis Whitman, a teacher and colleague of Wheeler's, asserted that the tendency to dichotomize into mechanism versus vitalism is destructive and confuses the important questions (1895; see Maienschein 1981).

When W. M. Wheeler invoked the organism as an appropriate metaphor for the study of social insects in 1911, what views was it designed to replace, and what choices were implicit for the 1911 reader that would situate Wheeler's theory? Similarly, when Wilson and Sober or Seeley defend, with some qualifications, the superorganism revival, what are the current distinct alternatives among which we are to prefer this approach? The metaphor appears to espouse the same transfer of structures from the primary domain of application of cellular organization of the body to the secondary domain of division of labor and specialization in the functional organization of societies. However, though the metaphorical word used may be the same, each may well be invoking a completely different meaning.

Let's look briefly at Wheeler (see Evans and Evans 1970). Wheeler began work in biology as a taxonomist and developmental biologist, publishing his first papers cataloging flora and fauna in his native Milwaukee in the late 1880s. He took his doctoral degree at Clark University in 1892, writing a thesis on insect embryology. He immediately took a position in the new University of Chicago, both under the influence of C. O. Whitman. He spent the next academic year in Europe, where he divided his time evenly between Theodor Boveri's lab in Wurtzburg and the zoological station in Naples. Wheeler was also a "regular" at the Woods Hole Marine Biology Laboratory in its earliest days. In 1899 Wheeler took a job and moved to University of Texas at Austin where he was to fall in love with the study of social insects, and with ants in particular. In Texas, Wheeler found himself virtually surrounded by numerous unidentified species of ant. He was to devote most of the rest of his career – academic positions at the American Museum of Natural History and a long-term association with Harvard University – to the investigation of social insects. Indeed, his observations are still now cited as primary resources, while his theoretical and conceptual frameworks have been less long-lived.

Wheeler's superorganism theory was proposed to explain what was characteristic of social insects. Wheeler saw this conceptual schema as a corrective to Darwinism because "... the struggle for existence is not more than half the truth.... To us it is clear that an equally pervasive and fundamental innate peculiarity of organisms is their tendency to cooperation or 'mutual aid'" (1923: 3).

The appeal to "mutual aid" is of course to Kropotkin. It is worth investigating why it was that the mutual aid or cooperation metaphors employed by Kropotkin would have found resonating voice in the United States in the 1920s, in lieu of Darwin's "struggle for existence" metaphor. Daniel Todes suggests that Darwin's "struggle for existence" competitive individualist metaphor reflects the "unsurprising fact that he [Darwin] shared the ideological outlook of his class, his circle and family and that such language might identify the author as bourgeois Malthusian, or perhaps, typically British" (1989: 13). Todes goes on to argue that Russia, the context giving rise to Kropotkin's "mutual aid" metaphor, was a land suffering not from population pressure, but rather from the harsh elements against which people had to work collectively to survive. Perhaps the United States of the late nineteenth and early twentieth century, and especially the Texas landscape where Wheeler developed his devotion to the study of social insects, was more geographically and demographically like Kropotkin's Siberia than Darwin's London. However, ecological determinism is a strong thesis, and there is little evidence to suggest it played a major role for Wheeler's superorganic metaphor.

Wheeler defined the concept of an organism, which would form the backbone of his theory of superorganisms, to include three features: organization for nutrition, reproduction, and protection.

> An organism is a complete, definitely coordinated and therefore individualized system of activities, which are primarily directed to obtaining and assimilating substances from an environment, to producing other similar systems, known as offspring, and to protecting the system itself and usually also its offspring from disturbances emanating from the environment. The three fundamental activities enumerated in this definition, namely nutrition, reproduction, and protection, seem to have their inception in what we know, from exclusively subjective experience, as feelings of hunger, affection, and fear, respectively. (1911: 5)

Wheeler argued that an animal colony is properly identified as an organism and is not merely an analog of one because it displays the following three features: (1) individuality: it behaves as a unitary whole, has identity in space, resists both dissolution and fusion with other substances; (2) duality of composition: it displays the Weismannian division of germ plasm and soma; and (3) ontogenetic and phylogenetic development.

21

Wheeler's superorganism model provided him with a framework to describe and explain the observed cooperation of individual insects. One can begin to define the boundaries of Wheeler's superorganism metaphor by describing what it was meant to exclude. First, he seems to be rejecting a narrow reading of Darwin. It was narrow by emphasizing competition and struggle and ignoring cooperation and mutualism. Darwinian competition was between individuals, and most severe between individuals of the same species. In this sense Wheeler can be viewed as anti-individualistic when competition excludes cooperation. Of course, Darwin also was prompted by the phenomena of sterile insects to entertain ideas of individual sacrifice to group benefit. "If such insects had been social, and it had been profitable to the community that a number should have been annually born capable of work, but incapable of procreation, I can see no very great difficulty in this being effected by natural selection" (1859: 236). Appeal to cooperation, at least in the case of social insects, was not an instance of anti-Darwinism, but rather opposed only to a narrow reading of Darwin.

This characterization of Wheeler is further supported by his defense of Darwinism in arguments with Father Erich Wasmann, a Jesuit living in Holland. Wasmann studied insect parasitism and guest species, one of Wheeler's favorite subjects, and was an outspoken anti-Darwinist who thought natural selection was inadequate for explaining the relationships he observed. His solution was to propose new instincts and new forces in its stead. Wheeler did not reject a role for natural selection; rather, he wished to expand Darwinism to include nonindividualistic perspectives as well.

Wheeler has been accused of another brand of anti-Darwinism, namely, of being a "crypto-vitalist" (Ghiselin 1974). He merits this label by his academic association with both the University of Chicago and Harvard University, which, according to Ghiselin, housed biologists, philosophers, and social theorists "who often explicitly denied that they were vitalists, but whose positions amounted to much the same thing" (1974: 30). This accusation is the result, we believe, of Ghiselin's strict dichotomization of views of the period. It seems that for him any view that is not individualistic, mechanistic, and neo-Darwinian in the sense of the impending synthesis falls into the category of vitalism. But this is unfair to Wheeler. Although he was attracted initially to Bergson, he later departed from vitalistic views, remarking that "the resort to such metaphysical agencies (as elan vital and others) has been shown to be worse than useless in our dealings with the inorganic world and it is difficult to see how they can be of any greater service in understanding the organic" (1926, quoted in Evans and Evans 1970: 224). In 1928 he referred to elan vital and entelechy as "little more than fetishes" (quoted in ibid.). Perhaps his most colorful rejection of vitalism appeared in 1911 in reference to Driesch's entelechy:

> His angel child . . . comes, to be sure, of most distinguished antecedents, having been mothered by the Platonic idea, fathered by the Kantian Ding-an-sich, suckled at the breast of the scholastic *forma substantialis* . . . but nevertheless, I believe that we ought not to let it play about in our laboratories, not because it would occupy any space or interfere with our apparatus, but because it might distract us from the serious work in hand. I am quite willing to see it spanked and sent back to the metaphysical household. (quoted in ibid.: 255)

Though eschewing association with the vitalists, Wheeler also rejected the genetic reductionism born of the synthesis of Weismann, Mendel, and population genetics that makes external selection the only relevant cause for explaining biological diversity, adaptation, and evolution. Internal organization and development were left out of the synthesis, and Wheeler's early work on embryology may have predisposed him to reject a Darwinism that ignored development. Writing in reference to the "exquisitely adapted" specializations in termites and ants, he says:

> And it strains our credulity to be told that such forms arise either from peculiar genes popping out of nowhere into the germ plasm or develop gradually under the guidance of natural selection from forms which, so far as I can see, must have an equal or even greater survival value. When we encounter such impasses as the foregoing, instead of embracing the Aristotelian Entelecheia . . . or joining the apostles of the survival of the fittest and forever croaking "natural selection!", it is surely more commendable to sit down in the laboratory or in the field and say nothing but "ignoramus" till we have made a much more exhaustive behavioristic and physiological investigation of the phenomena. (1928; quoted in ibid.: 261–262)

Nevertheless, Wheeler did not reject Darwinism entirely. He supported a broader interpretation of Darwinian theory, one that could redress the lacunae of developmental, internal processes made invisible by the genetics of his time.

In summary, Wheeler invoked the organism metaphor for social insects in the form of his superorganism theory against a strictly external selectionist interpretation of Darwin. This put him in the company of many Darwinian critics of the time, those concerned with internal processes, such as embryogenesis and other developmental considerations, who found that individual competition in response to external environmental conditions was insufficient to ground the kind of phenomena they were describing. Wheeler was, after all, a taxonomist constantly reminded of the variations distinguishing the thousands of species of ants, bees, wasps, and termites and an embryologist who saw complexity developing in the lifetime of the organism. While competition between genetic individuals may not have been sufficient, neither did it require embracing a vital force, such as Dreisch's entelechy, Maeterlinck's spirit of the hive, or Bergson's elan vital, an agency existing separately from matter that directs, selects, or arranges the structure of matter that constitutes "life." Wheeler seems to have occupied a position that corresponds to what Tim Lenoir (1989) has described as "vital materialism," or a midpoint in Jane Maienschein's suggested continuum between mechanistic materialism and vitalism (1981). Wheeler vehemently opposed vitalism, while at the same time rejecting the complete reduction of biology to a form of physics.

The "Revived" Superorganism

Wilson and Sober (1989) use the superorganism concept to promote a hierarchical theory of selection. Their interpretation of Darwinian natural selection logically entails the rejection of privileging the individual organism as the only level at which selection can operate. Again, one can locate their position in contrast to a narrow interpretation of Darwinism, the latter being what S. J. Gould has called the hardening of the synthesis (Gould 1983). For them, higher levels of organization – such as groups, species, or superorganisms – are reasonable candidates for selection. Furthermore, Wilson and Sober argue that not only is it theoretically possible to have such levels of selection, but in fact the conditions for their realization are not overly restrictive (contra the early objections to group selection; cf. Wade 1978) and, in fact, they do exist in nature. However, it is ironic that the strategy they use to reject the hegemony of a single individual level of selection has features that are similarly restrictive. Just as the individual organism prejudice tends to privilege a single level as the only one relevant for selection, blinding biologists to the other levels, Wilson and Sober's superorganism theory privileges a particular form of functional organization at the group level, blinding one to diversity in this realm. The individual single-level theory they reject by invoking the superorganism metaphor is merely replaced by a group single-level theory, thereby obscuring the multiple levels at which selection can and does operate in social insect societies.

For Wilson and Sober, a superorganism is "a collection of single creatures that together possess the functional organization implicit in the formal definition of organism" (1989: 339). They adopt the Random House dictionary definition of an organism to be "a form of life composed of mutually dependent parts that maintain various vital processes" (Wilson and Sober 1989: 339). While this formal definition is consistent with Wheeler's articulation of three essential activities of nutrition, reproduction, and protection – that is, the various vital processes – Wilson and Sober's use of functional organization is further restricted to organization around reproduction. They summarize their superorganism model as follows:

(i)   A population is subdivided into a number of groups. . . .
(ii) Groups vary in properties that affect the number of dispersing progeny (group fitness).
(iii) Variation in group fitness is caused by underlying genetic variation that is heritable. . . .
(iv) No differences exist in the fitness of individuals within groups. (1989: 339)

Wilson and Sober (1989: 340, 343) claim that functional organization is the key to identifying at which level natural selection is operating. Functional organization is that which allows individual organisms or superorganisms to successfully survive and reproduce. When the four conditions are met, they claim that natural selection will endow groups with the same properties of functional organization that we normally associate with individual organisms (1989: 342). Their argument is as follows. Since natural selection can act both within and between groups of individuals, be they organisms or superorganisms, there are three possibilities. First, within-group selection overwhelms between-group selection, and the result is a collection of individuals. Second, between-group selection overwhelms within-group selection, and the result is an organism in the formal sense. Third, when neither level wins, the result is part collection and part (super)organism.

Implicit in their development of a superorganism is a particular form of functional organization of the parts of the paradigmatic individual organisms, namely, a Weismannian preformationist developmental schema. On this view there is a sharp division of germ plasm from soma, and the germ line is sequestered so early in development that there is no opportunity for competition between somatic cells to have evolutionary consequences. The Weismannian legacy makes variation during the development or experience of the parts (soma or workers) irrelevant to evolution because it entails the complete, early separation of germ cells from somatic cells. Weismann's theory, requiring the rejection of Lamarckianism, was a cornerstone of the modern synthesis.

However, Buss (1987: 20) has convincingly argued that this very ideal image of the functional organization of cells in an organism fails to be approximated by most taxa or throughout most of geological time. Buss's list of all the taxa displays how rarely the Weismannian ideal is met in the primary domain of the biology of individual organisms. So, too, is the idealized Weismannian superorganism rarely approximated by the range of organization found in social insects.

The Variety of Functional Organization in Social Insects

There are more things in heaven and earth, Horatio, than are dreamt of in your philosophy. (Shakespeare, *Hamlet*, first quarto, lines 607–608)

The Wilson-Sober model of superorganisms requires that there be no differential "fitness" between units at levels lower than the colony or between levels (their condition iv). Differential fitness is differential survival and reproduction for them, the conditions required for there to be evolutionary consequences of selection at a given level. However, competition at levels lower than the colony can occur among individual colony-mates, among males, or even among genes, for example, with meiotic drive or segregation distortion genes. Competition can even occur between workers and queens over determining the colony sex ratio. Any of these would make the colony something less than a superorganism. The superorganism ideal may be achieved by either eliminating all genetic variation among the subcolony units, and thereby removing a necessary condition for selection to operate at that level, or by suppressing all competition among genetically variant subcolony units at all levels by means of mechanisms, such as meiosis, for making such variation effect no change in the distribution of traits in offspring.

Rather than produce a complete table of social insects, listing those that do and those that do not qualify under the stringent condition of no variation in fitness at below colony levels, we instead point to some of the sources of within-colony competition that result in most, if not all, social insects failing the criterion.

Haplodiploidy leads automatically to differential fitness among male members of societies. Males arise from unfertilized eggs laid by diploid females, queens, and workers. Recombination in females leads to a vast array of potential genomes embodied in egg gametes. If left unfertilized, the egg undergoes development and maturation and produces a haploid male individual. Each male then replicates the genome that originated it thousands or millions of times (there is no recombination) and changes the sex of the cell embodying the genome from female (egg) to male (spermatozoon). These identical male gametes are packaged within haploid individual organisms (males) that then

26

compete for resources within the nest, such as food and care from workers, and for mates outside the nest. Their competitive abilities are at least to some extent dependent on the gamete that produced them and, therefore, represent a form of gametic/individual selection unique to haplodiploid organisms. (This is different from what we normally think of as sperm competition because the competitive abilities of spermatozoa are probably mostly determined by the genotype of the male that produced them.) Elimination of competition at this level would require that the reproductive success of males be independent of their genotype, an unlikely condition.

Termites, on the other hand, are diploid and monogamous and do not have these particular difficulties. However, termite societies often have supplementary reproductives and not a single king and queen, probably resulting in within-colony variability in reproductive success during at least some stages of colony development (Wilson 1971; see discussion of polygyny below).

Polyandry is characteristic of many species of ants, wasps, and bees in which the queen of a colony mates with multiple males. This reproductive strategy can give rise to many different sources of within-colony competition. First, males must surely compete for matings with the queen on her nuptial flight, as they compete for resources within the colony during their early development and outside it while they prepare for reproduction. The queen mates during a few days of her life, then forms a colony and dispenses the stored sperm over her lifetime, which for honeybees can be several years. Thus, competition between males may reflect fitness differences for the ability to mate at all, for quantity of sperm deposited, for longevity of sperm for use during the queen's lifetime, and for the ability of sperm to fertilize queen eggs rather than worker eggs. There is evidence that sperm use by the queen is not random. Given that males differ genetically, such fitness differences, which obtain below the level of the colony, are likely to have evolutionary consequences.

In addition, given that multiple males father the "body" of female workers, different subfamilies are formed in a colony. (All workers share the same mother, but some will share the same father and differ from others in that respect.) In the social Hymenoptera, females develop into queens or workers on the basis of what they are fed. If workers that feed larvae can recognize those that are more closely related to them, that is, members of their own subfamily, then competition may arise among individuals of different subfamilies over which larvae are raised as queens versus workers. Competition can be eliminated if workers are unable to distinguish among potential queen larvae. Empirical work with honeybees has demonstrated that workers have the ability to discriminate among larvae under some experimental conditions

(Noonan 1986; Page and Erickson 1984; Page et al. 1989a; Visscher 1986; cf. Oldroyd et al. 1990 and Page and Robinson 1990).

To eliminate competition due to polyandry at these levels, the following unlikely conditions must be met:

1. Mates of queens contribute equal numbers of sperm;
2. Sperm will have equal viability and longevity;
3. Sperm are used randomly to fertilize eggs; and
4. Workers are unable to discriminate among larvae with different subfamily relationships.

These are very unlikely conditions.

In addition, polygyny, or multiple queens, is widespread in the termites, wasps, and ants (Wilson 1971) but is rare in bees (Michener 1975). Some ant colonies have literally thousands of functional queens. Here competition within the colony can arise through differential egg laying by the different queens. Also, like polyandry, it could result in competition among workers in the different subfamilies generated by the multiple queens. Competition among queens can be eliminated only by a lack of heritable variation in the production of reproductive offspring by the queens. This seems also to be an unlikely condition to be met.

Genotypic variability among individual organisms and, therefore, variability in reproductive success within insect societies may arise thorough polyandry, polygyny, and recombination. Larvae with different genotypes, for any of the above reasons, may compete to become reproductives. Larval competition may explain why it takes 21 days for a worker honeybee to develop from egg to adult, while it takes a queen only 16 days, even though she is twice the size of a worker. Or why the fate of a larva, with regard to its development into a queen or worker, is not determined until more than halfway through larval development, the only phase of development where immature stages feed and can be fated to become a queen.

Several virgin honeybee queens are simultaneously reared by colonies and fight to inherit the maternal nest after the queen mother leaves with an entourage of workers. The first queens to mature, that is, that have the shortest development time, have an advantage over those that are still developing because they sting and kill immature queens before they complete their development into adults.

Sometimes a colony loses its queen to disease, old age, or accident and raises a replacement by selecting a few young worker larvae from among thousands to raise as queens. Only those immatures that are not already fated developmentally to become workers can become queens. Those that become queens obviously have greater reproductive success than those that become workers, although one might argue that if as workers they raise only sisters related by three-quarters, then they have greater reproductive success than producing sons and daughters related by half. Even so, queens live a lot longer and produce far more reproductive offspring than workers care for in their short lives, so the reproductive success of being a queen, rather than a worker, may be quantitatively greater.

Egg-laying workers are common in societies of ants, wasps, and bees (Bourke 1988). They often occur simultaneously with one or more functional queens and even more often in colonies that are queenless. Studies of honeybees have shown unequal, but significant, reproductive success of laying workers in queen-right and queenless colonies (Page and Erickson 1988; Robinson et al. 1990; Visscher 1989; see below for more details). Some species of ants belonging to the genera *Monomorium*, *Pheidole*, *Pheidologeton*, and *Solenopsis* completely lack ovaries and therefore are exempt from any possible individual reproduction (Hölldobler and Wilson 1990). However, at least some member species of these genera are polygamous and/or polyandrous (Page 1986; Ross 1989).

Social insect colonies will fail to be superorganisms by failing to meet condition iv, the complete suppression of competition at levels internal to the organism, the extension of the Weismannian ideal model to social groups. There are good reasons to believe that there is reproductive competition among colony units at lower organizational levels. The most likely candidates to qualify as a Wilson and Sober superorganism would have a single queen that mates with a single male and workers that lack functional ovaries. Admittedly, some ant species fit this ideal; however, they represent only a small fraction of the different kinds of colony organization displayed by social insects. Insect societies are rich in diversity with respect to how they are functionally organized for reproduction, just as Buss has shown that individual organisms are similarly diverse in their internal functional organization. The social insect colonies, even when different in their reproductive organization, may be similar with respect to organization around defense or nutrition, two aspects of a superorganism that Wheeler emphasized and Wilson and Sober ignore. That these strong conditions are intended by Wilson and Sober is shown by their claims that "social insect colonies really do cease to be superorganisms, to the extent that natural selection operates within single colonies" (346), and "The essential criterion is absence of within-group selection, which may be accomplished either by creating genetically uniform groups, or by suppressing the differential reproduction of genetically diverse groups" (348).

Wilson and Sober promote the revival of the superorganism, but it is not Wheeler's metaphor that they invoke. The major differences are:

1. Wilson and Sober limit functional organization to organization around reproduction, ignoring the nutritional and protective organization that Wheeler included; and

2. While Wheeler found selective scenarios insufficient to explain superorganismic phenomena, Wilson and Sober make selection at the group level the defining characteristic.

The scientific context for Wilson and Sober, the contrast class in which to situate their superorganism, is the individualistic theories that "have dominated evolutionary biology for the last twenty years" (353). The superorganism is supposed to provide a "radical departure" from this narrow interpretation of Darwinism by acknowledging multiple levels of selection. They promote a hierarchical picture of natural selection, one not restricted to a single level, yet one logically consistent with the teachings of Darwin. In these regards, Wheeler and Wilson and Sober are similarly fighting their respective hegemonic, narrow interpretation of Darwinisn, offering radical departures that nevertheless do not step outside the bounds of what is "truly" Darwinian. Wilson and Sober's superorganism, however, depends on the Weismannian ideal organism and not the contemporary picture of a plurality of organismic organization. The horizontal transfer of content with respect to the organization of the whole from its parts seems to be in their case oblique. The Weismannian organism when elevated to the Wilson and Sober superorganism contravenes their very goal of supporting a broad and hierarchical Darwinian theory. The Wilson and Sober superorganism makes invisible the diverse array of truly hierarchical organization found in social insects on which hierarchical selection can operate. Their agenda is to rid biology of the myopic picture of selection operating only at the individual organism level. Their explicit reasons for reviving the superorganism model concern the inconsistency of adopting a theoretical framework that focuses only at the individual level. And while in fact they admit within-organism selection (in the primary context of the metaphor) as well as within-colony selection (in the secondary context of the metaphor), their overly restrictive model of the superorganism obscures these admitted facts.

By exploring the transfer of content and contexts of the original superorganism metaphor for social insects developed by William Morton Wheeler and comparing it with the "revived" superorganism of Wilson and Sober, the complex ways in which metaphors acquire their meanings begin to be clarified. The appeal to ancestral authorities to inspire allegiance to a theoretical construct may have rhetorical success, but only by virtue of ignoring the differences that made appeal to the metaphor salient in its own historical and scientific milieu. The Wilson and Sober superorganism's credentials as a "revived" form of Wheeler's original metaphor can be judged only by such a comparison. And as argued above, their model and that of Wheeler are substantially different.

Whether the superorganism metaphor and model proposed by Wilson and Sober should be accepted, regardless of its pedigree, depends on the content and promise of such a conceptual framework. The model's content is understood by means of the horizontal transfer of structure from the primary context of application, the organism, to its new domain, social insects. On that score we have argued that, ironically, the model obscures our vision of just the sort of variability Wilson and Sober wish to highlight, and it does this by transferring the Weismannian ideal structure of the organism to define the structure of the superorganism.

Wilson and Sober give us an idiosyncratic view of insect societies that may make invisible the diverse array of truly hierarchical organization on which hierarchical selection can operate. They are rightly concerned to rid biology of single-level theories in order to acknowledge multiple levels of selection and emphasize the fact that population structure matters to evolutionary processes, that is, that forming groups can affect the selection process and its consequences. Hence, a perspective that sees only a population of individuals will not be able to take such levels into account. We believe that a better route to encouraging this revision of evolutionary theory is in terms of a complex systems model that admits all the variations of functional organization. This would be more successful in addressing multiple levels of selection than the rigid, idiosyncratic superorganism model.

Insect Colonies as Dynamical Complex Systems

Even if we granted for the moment that social insects approximate the ideal, stringent criteria of the Wilson-Sober model of superorganisms, the dynamically changing status of such an organism is not represented in their theory. Not only are candidates for superorganisms extremely rare, but the superorganism status of a colony is not stable and may change dramatically with colony development. For instance, colonies of most monogynous species of social Hymenoptera change their reproductive structure when they lose their queen (see Bourke 1988). Workers undergo individual development where their ovaries become mature and they begin laying unfertilized eggs (or fertilized, if the workers are capable of mating). This can be demonstrated in the

31

honeybee, where there can be at least three identifiable colonies, depending on the reproductive status of workers.

State 1: Queenright Colonies. Queenright colonies have a functional queen. Nearly all reproductive individuals are progeny of the queen. This is the case emphasized by Seeley (1989) in his description of honeybee societies as superorganisms.

In at least some colonies, workers lay some unfertilized eggs that develop into males (Page and Erickson 1988; Visscher 1989). Page and Erickson (1988) reported one colony where all males produced were probably progeny of laying workers. Ratnieks and Visscher (1989) suggest that conflict occurs among workers within colonies, resulting in the cannibalism of most worker-laid eggs.

State 2: Temporarily Queenless Colonies. The first functional response to the removal of queens is the desuppression of queen rearing. Workers construct queen cells and raise a new replacement queen. If the colony successfully raises a queen and the queen successfully mates, then the colony returns to state 1. If the new queen fails to mate successfully, but survives, the colony returns to a degenerative state 1 colony where only drones are produced but they are the sons of the virgin queen. If the queen fails to survive or no queens are successfully raised, then the colony goes to state 3.

State 3: Permanently Queenless Colonies. Suppression of worker ovary development and egg laying is usually lifted after the loss of the queen, and all of the immature workers complete development and become adults (see Robinson et. al. 1990 for review). After workers begin the transition from state 2 to state 3, they can be attracted to one of two alternative states:

State 3a: Here, one or more workers produce queen pheromone and suppress egg laying of other workers (see Robinson et al. 1990). These "false queens" then become the principal germ line for the workers. Colonies in this state appear to produce far fewer reproductives than the alternative state 3b.

There is an alternative trajectory for state 3a. Some races of bees produce females parthenogenetically at high frequency (all do it at some very low frequency). Those capable of high-frequency female production manage to maintain the integrity of the state 3a condition for several weeks or months and then raise a new reproductive queen and return to state 1 (Anderson 1963).

State 3b: This state is characterized by a progressive increase in egg laying by workers and appears to have some special functional properties, including: egg laying preferentially in drone-sized cells (this is important because drone larvae reared in worker brood cells develop into much smaller, and presumably less reproductively competitive, drones than those raised in special, larger-sized drone brood cells), egg and larval cannibalism, and a synchronous onset of oviposition by laying workers (Page and Erickson 1988; Robinson et al. 1990). Egg-eating and egg-laying behavior are distributed nonrandomly among workers belonging to different subfamilies within colonies. The net result of differential egg laying and egg and larval cannibalism is differential reproductive success among colony members.

To summarize our argument so far, we have presented evidence for why the Weismannian superorganism model proposed by D. S. Wilson and Sober fails to describe most social insects. In addition, we have suggested that even if we relax their constraints on the defining functional organization of superorganisms so that existing social insect societies would more closely approximate it, the pure superorganism state remains an unstable one during the lifetime of the colony. We now argue that in addition to the organizational and dynamical variability that is hidden by a superorganism theory, there is the possibility of spontaneous origins of order at the colony level. That is, the Wilson-Sober model lends itself to a type of selectionism, albeit at the group level, which ignores the possibility of nonselectionist developmental explanations of colony organization.[2]

Functional or Apparently Functional?

A snapshot look at a complex system shows what appears to be the intricate, coordinated behavior of parts of the system in a seemingly goal-directed manner. Such coordination in living systems has led some to postulate an elan vital or entelechy as an invisible hand directing the coordination of the parts. For social insects, a spirit of the hive (Maeterlinck 1912) was postulated, which, like the soul of the body, was held responsible for an order that could not be the result of mere mechanical means.

These vitalistic interpretations long have been rejected by the biological community and replaced by mechanistic forces operating on living organisms. Natural selection, operating on heritable variation among individuals, can put into place features that seem consciously designed to solve the problems of existence that any organism faces: nutrition, reproduction, and protection. The primary role of natural selection as the only source of order or organization has been successfully challenged (Gould and Lewontin 1979; Kauffman 1984; Page and Mitchell 1991). Wilson and Sober's (1989) model of a superorganism theory and Seeley's (1989) use of it suffers from the same selectionist

---

[2] E. O. Wilson's superorganism theory does not suffer from the same sorts of restrictions. The major target of his theory is to account for colony development, that is, to provide a theory of sociogenesis.

bias that has characterized single-level organism theories of selection in the past. That is, they take the observation of apparent functional organization to be evidence of a history of group-level selection. This bias is the result of their account of functional organization.

Functional ascription has played two roles in biological explanation. One has been as a purely descriptive account of the interrelatedness of parts of a system and their contribution to the system's capabilities (Boorse 1976; Cummins 1975; see Sec. 4.3 below). The other identifies functions with adaptations (Millikan 1984;Wright 1976). On the second view, identifying a feature as functional entails the claim that the feature had been selected in the past for its contribution to the system's survival and reproduction and has evolved as a result of that selection (Mitchell 1989). It is unclear which of these interpretations should be attributed to D. S. Wilson and Sober. On the one hand, they attempt to revive Wheeler's superorganism theory, which was clearly based on functional organization of the first type. Indeed, they say that functional organization is that which allows individuals to successfully survive and reproduce (1989: 340), a claim about the proximate contribution of parts to contemporary goals. Yet they also say that functional organization is the accumulation of properties that enhance relative survival and reproduction of individuals (343), a claim about evolutionary history. This blurring of the use of the concept of functional organization allows them to make the transition from an observation of the contemporary functional organization of a superorganism that appears to contribute to survival and reproduction to the conclusion that such functional organization must be the result of natural selection in the past. Their argument then uses the restrictive Weismannian ideal form of reproductive organization to further conclude that the selection must have been at the group level.

Antiadaptationist arguments have pointed to two ways in which evolutionary histories may diverge from the paradigmatic case of direct selection of a trait for its consequences on fitness. On the one hand, selection may be indirect, that is, operating on one trait and at the same time giving rise to evolutionary change of another trait by means of the developmental relationships between two traits. On the other hand, the evolutionary change in a trait may be the result not of selection, direct or indirect, but rather the spontaneous emergence of order.

Direct versus Indirect Selection

Evolutionary histories of given traits may not include direct selection or may be a mixture of episodes of selection operating at more than one level. The

Wilson-Sober superorganism model's identification of functional organization with group-level selection fails when the organized results are really emergent properties that result from individual level selection. West-Eberhard (1981) argues that many of the features of insect societies that appear cooperative in nature, and hence likely to arise through higher levels of selection, are easily explained as historical contingencies resulting from competition among alternative reproductive strategies of individuals. She argues that queen-worker division of reproductive labor and even worker age-polyethism, two of the presumably primary functional components of eusocial insect societies, may in fact have evolved through these competitive interactions. Thus, the group selectionist perspective may fail to see the truly hierarchical structure of selection operating.[3]

Selection versus Spontaneous Order

The Wilson-Sober assumption that all apparent functional organization is the result of a selection history can also be challenged. The three fundamental properties of insect societies that are regarded as the reason for their great ecological success – division of labor, mass action responses, and social homeostasis (Wilson 1985b) – may, in fact, be self-organized properties of groups and not the result of selection at any level. With simple computer simulation models, Page and Mitchell (1991, 1998) demonstrated that an ensemble of "computerized bees" exhibit a division of labor similar to that observed for social insects with only two requirements: First, they are tolerant and interact within a common environment, and second, the individual group members have different thresholds of response to sets of stimuli that elicit task-performing behavior. Individuals can have different response thresholds because of differences in individual experiences and learning; because they have different genotypes that set, or limit, thresholds of response; or because they are in different states of behavioral development. Age and experiential changes in behavior are characteristics that are not unique to social insects. In these cases, a selectionist perspective veils the intriguing potential for insect societies to display the spontaneous order that has been proposed for other biological systems (Kauffman 1984).

---

[3] This question about the actual level of selection (individual or group) is not equivalent to the debate concerning the unit of selection (replicator or interactor). The issue at hand concerns which causal interaction processes were involved in the evolution of a given trait, not whether any such interaction can be represented in genic terms. Wilson and Sober are well aware of this important distinction, but it is worth repeating to avoid unnecessary confusion.

Toward a Dynamical Model for Hierarchically Organized Complex Systems

Given the variety found in the empirical record of social insects and our arguments concerning the theoretical blinders that the superorganism theory entails, we wish to promote in its place a theory that avoids the pitfalls and adequately explains the phenomena. We suggest that such a theory have the following features.

1. Hierarchical selection (see Wade 1980): a theoretical model that embraces the multiplicity of levels at which selection can and does act and does not preclude simultaneous selection at more than one level (Wade 1980; see also Lewontin and Dunn 1960; Vrba 1989). Wheeler (1911, 1928) and E. O. Wilson (1985a) clearly allow for hierarchical selection to operate on their superorganisms, while D. S. Wilson and Sober (1989) allow selection only at the level of the group.

2. An ontological framework that corresponds to this hierarchy, namely, a focus on the properties of individual components of ensembles as well as the properties of the whole. Focusing on superorganism-level phenomena alone restricts our ability to understand the actual mechanisms involved in building higher-level phenotypes. Natural selection at the group level must change properties of component parts (individual organisms) in order to achieve higher-level functional properties. Gene frequency changes resulting from selection are a result of the substitution of alternative alleles at gene loci residing in the cells of individual organisms. The individual organisms define, in a sense, the parameter space on which higher levels of selection can operate and constrain the possible group-level phenotypes.

3. Multiplicity of explanatory strategies: a model that does not entail selectionism and hence allows for spontaneous order to emerge from interactions of component parts and explain group dynamical properties. The use of functional organization to simultaneously describe the current contribution of a part of a system to the system's survival and reproduction and to pick out an arrangement that is the result of a natural selection history makes nonselectionist explanations of functional order impossible. We need a descriptive language that does not bias the outcome of investigation into the developmental dynamics or selection histories of a given system.

4. This model should not be based on a restrictive Weismannian paradigm and instead should recognize multiple forms of functional reproductive organization and allow the conflict between levels of selection and among component parts and its incomplete resolution.

Conclusion

The superorganism metaphor and its more sophisticated relative, the theory of superorganisms, are used for a variety of purposes by their proponents. For Wheeler, the superorganism model provided a mechanistic framework to account for the observed cooperation of individual insects. He proposed it both as an alternative to the scientifically suspect invocation of metaphysical spirits and entelechies and as an alternative to a narrow reading of Darwinism that presumes a world filled with conflict among individual organisms. E. O. Wilson revived the superorganism as a developmental framework in order to model sociogenesis of social insects akin to the morphogenesis of individual organisms. D. S. Wilson and Sober and Seeley hope to make the superorganism theory do the bidding of a hierarchical theory of selection as an alternative to a single-level individual selectionistic framework.

In this chapter we have focused on the shortcomings of the Wilson-Sober model of the superorganism, not as an argument against the promotion of a hierarchical model of selection, but rather in its favor. We support a pluralism of explanatory strategies to apply to social insects, including cooperative, developmental, and hierarchical selection frameworks. However, because of the multiple purposes for which the superorganism concept has been used, and hence its equivocal reference in the contemporary revival, we believe such goals can best be achieved by developing an explanatory model in other terms.

3

Dynamic Complexity

Understanding the dynamical mechanisms by which complex organization is built can provide important insights into the evolution of complexity. Indeed, adopting an "adaptationist" view of the biological world, where all atomized structures are thought to be the result of direct selection over variation for fitness consequences, systematically ignores the role that development of structure plays in evolutionary history. In this chapter, written by Robert E. Page and Sandra Mitchell,[1] we attempt to take seriously the developmentalist challenge. That is, if we can see *how* division of labor and other features of social insect colonies might arise, we can better understand which features of a complex phenotype were available for variation and possible selection in the evolution of sociality.

Self-organization refers to a family of agent-based models for generating "order" at a higher level from the interaction of components at a lower level without requiring the resulting structure be coded for in genetic blueprints or be solely a result of centralized control structures. The models that we have developed were inspired by Stuart Kauffman's N-K Boolean networks (Kauffman 1993). We imbued the individual computer bees with characteristics found in solitary bees and ran simulations to determine which features would result from the mere interaction of such individuals. The results were interesting. Division of labor emerged "spontaneously" from the self-organizing dynamics of our model.

What this implies is *not* that natural selection did not shape the features found in contemporary social insect societies, but rather that at least some

---

[1] This chapter is a further development of a joint paper by Robert E. Page, Jr., and Sandra D. Mitchell, published in A. Fine, M. Forbes, and L. Wessels, eds., *PSA 1990*, vol. 2 (East Lansing, Michigan: Philosophy of Science Association, 1991), pp. 289–298. R. E. Page, Jr., and S. D. Mitchell, "Self Organization and the Evolution of Division of Labor," *Apidologie* 29 (1998): 101–120, reprinted by permission.

aspects of those features might well be the result of complex dynamics of the development of insect colonies. That is, natural selection could well have tuned the features that arose from self-organization to render variations of these structures more adapted to particular environments, but it may not be the whole story. Indeed, recent empirical work (Fewell and Page 1999) supports these conclusions.

In the long-standing debate about the interaction of developmental and evolutionary processes in generating the phenotypes of organisms we observe, these models suggest that features of sociality are the product of the ontogeny of a colony of individual insects. That is, division of labor itself may well not have been the consequence of natural selection sorting between colonies with such an arrangement of workers and colonies without such an arrangement. Rather, any group of living insects with the properties of solitary insects will automatically divide their labor in response to internal and external stimuli. Natural selection operates on the variations that still persist in aspects of division of labor, turning this structure to the environment in which social insects live, but it is unlikely that it was solely responsible for its appearance and maintenance.

3.1. THE EVOLUTION OF DIVISION OF LABOR

Robert E. Page, Jr., and Sandra Mitchell

Introduction

The main features of insect societies that are believed to be adaptations responsible for their tremendous ecological success are (Oster and Wilson 1978; Wilson 1985a, 1985b): *division of labor*: between reproductives and workers, and a further division among workers that is often, perhaps usually, based on age and/or anatomical differences; *specialization*: some individuals perform some tasks with a significantly greater frequency than do other individuals; *homeostasis*: colonies regulate internal conditions, such as food stores, temperature, and humidity; *plasticity* and *resiliency*: colonies are able to change the numbers of workers engaged in different tasks in response to changing internal and external colony environments; and *mass action responses*: colonies are able to mobilize large numbers of workers for specific emergency needs.[2]

[2] This work was funded by National Science Foundation Grants BNS-8719283 and BNS-9096139 to R. E. Page and a Chancellor's Summer Faculty Fellowship, University of California at San Diego to S. D. Mitchell. S. D. Mitchell also thanks the Population Biology Center, University of California, Davis, for partial support.

Much of the classical and contemporary research in insect sociobiology has been dedicated to studying these phenomena with the assumption that they represent colony-level functional adaptations. To identify a given trait as an adaptation is to imply that it has evolved by means of natural selection on available variants (Gould and Vrba 1982; Harvey and Pagel 1991; Lewontin 1978; Mitchell 1987b; Sober 1984c; Williams 1966). For example, if division of labor and specialization are adaptations, then we assume a prior competition between colonies, some with a heritable division of labor and specialization, some without, and a colony-level selection process acting to preserve those traits through differential colony reproduction. However, if all groups of insects demonstrated these colony-level phenomena as a consequence of group living, then these characteristics are not themselves adaptations, even though their specific features may be. For instance, as we argue below, task specialization may be an inescapable property of groups of individuals that tolerate each other and share a nest, but having specific caste ratios may be adaptations to specific environments (Oster and Wilson 1978).

The historical definition of adaptation we, and most biologists and philosophers, use has recently been criticized generally, and our arguments about division of labor specifically (Page and Mitchell 1991), by Reeve and Sherman (1993). They promote instead a nonhistorical account of adaptation as "a phenotypic variant that results in the highest fitness among a specified set of variants in a given environment" (1993: 9). The major and, we believe, devastating drawback of their definition is the complete disconnection between adaptation and evolution entailed by it. For a trait to be an adaptation, on their account, says nothing at all about its relation to traits in past or future populations. This is in contrast to Darwin's notion of natural selection acting on variation as the mechanism for evolutionary change producing, and hence explaining, the adaptation of traits to their environment. Our 1991 argument that division of labor could have arisen by self-organization implied that there would have been no variance between protosocial groups for the fundamental aspects of division of labor and hence no natural selection for those features. This view was criticized by Reeve and Sherman (1993). They mistakenly alleged that we see self-organization and selection as alternative, rather than complementary, explanations (as do Bourke and Franks 1995: 421; see also Mitchell 2002c). This reading of our paper is difficult to understand because we elaborated in that paper, and again below, the ways in which once a crude division of labor has emerged, natural selection can then operate on the variations in features of this colony phenotype to preserve those variations that are more fit.

Here, using a Boolean network modeling approach (Gelfand and Walker 1984; Kauffman 1984, 1993; Page and Mitchell 1991), we demonstrate how extremely ordered group behavior can emerge from random aggregations of individuals. Our models require only that individuals have response thresholds associated with stimuli that result in behavior that in turn affects the current levels of stimuli. These are the very mechanisms believed to result in functional behavior of all animals, solitary and social. However, when living in groups, the actions of some individuals in response to common stimuli change the stimulus environment of others and, in turn, influence the behavior of other group mates (see also Bonabeau et al. 1996; Page 1997). As a consequence, some features of insect societies are inescapable outcomes of interactions of individuals. We suggest how colony-level selection may change the organizational structure of colonies by acting on characters that affect system-level parameters.

The Model

The model we construct here is analogous to a protosociety of insects. We later use honeybees as the exemplar organism because we make a transition from the simple organizational patterns of our models into exploring how more complex social organization may evolve. We assume that solitary insects suddenly become tolerant of each other and share a nest. However, there is no reason to assume that nontolerance is the ancestral condition, because many insects develop together in common nests and then disperse after becoming adults. We simply want the initial conditions to assume that there is no prior history of group living that could have shaped specific patterns of social interactions.

Assume that individuals are analogous to elements (or nodes) of a binary switching network (Gelfand and Walker 1984; see Fig. 3.1). Individual elements (insects) can be either *on* (1) or *off* (0) with respect to their likelihood of performing a particular behavioral act. Also assume that the number of elements in the network, N, represents all individuals that belong to a particular behavioral state that are behaviorally competent to perform that task. K, the connectance (Gelfand and Walker 1984) of individual network elements, is large, close to or equal to N, with elements receiving directed inputs from and sending outputs to all other elements. In other words, all $N - 1$ elements (workers) receive information about themselves and the state status of all $N - 1$ other elements in the network. Individual input lines to elements are *on* or *off* according to the state of the element that initiates it. Each element is assigned a switching function (a decision rule) that determines whether the

Table 3.1. *All $2^{2^k} = 16$ Boolean Functions for Binary Elements with* **K** $= 2$ *Inputs*[a]

Inputs		Boolean functions															
X_1	X_2	1	2	3	4	5	6	7	8	9	10	11	12	13	14	15	16
0	0	0	1	0	0	0	1	1	1	0	0	0	1	1	1	0	1
0	1	0	0	1	0	0	1	0	0	1	1	0	1	0	1	1	1
1	0	0	0	0	1	0	0	1	0	1	0	1	1	1	0	1	1
1	1	0	0	0	0	1	0	0	1	0	1	1	0	1	1	1	1

[a] The number of Boolean functions increases exponentially with K; therefore, we present the set of Boolean functions for the simple case of $K = 2$. Individuals realizing function 1 (f_1) are *off* regardless of the values on their inputs, X_1 and X_2, while those with f_{16} are always *on*. f_2 is a threshold function that returns a 1 (*on*) only for the case where all inputs, t, are *off*. f_5 is *on* for $t > 1$, f_{12} for $t < 2$, and f_{15} for $t > 0$.

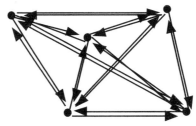

Figure 3.1. A directed arrow graph for the case $K = N - 1$. Each element (node) has an input arrow from all other $N - 1$ elements and an output arrow to all $N - 1$ elements. Each arrow carries an *on* (1) or *off* (0) value to the element at the arrow's head. The value carried by the arrow is the current value (*on* or *off*) of the element that initiates it (located at the tail). Each element is *on* or *off* as a consequence of the number of *on* or *off* inputs to it. For the $K = N$ models presented, each element has an arrow to itself.

element is in an *on* or *off* state according to the numbers of *on* and *off* inputs coming to it from all other elements. Thus, each element has a threshold of response.

Individual threshold functions, f_i, are assigned randomly to elements from the set $\{F'\}$, which is a subset of the set of all Boolean switching functions, $\{F\}$, that has as its members all and only those functions that switch the element *on* or *off* when the number of *on* or *off* inputs exceed some specific value (Table 3.1). During simulations of the model, network elements check the values on each of their inputs, evaluate, then respond by turning *on* or *off*. Assume for this model that individual elements are switched *off* when the number of *on* inputs is equal to or exceeds some given threshold value, and switched *on* when the number of *on* inputs falls below that threshold. Individual elements can check their inputs and respond by switching *on* or *off* in three different ways: (1) *simultaneously*, where all individuals check their inputs at the same time and respond together; (2) *sequentially*, like they are queued up at a turnstile and check their inputs and respond individually in a repeated, sequential order; or (3) *randomly*, where each element is sampled individually and randomly, checks its inputs, and responds.

An example of how this model may represent a honeybee society is to assume that individual workers (elements) receive information of the state (*on* or *off*) status of other workers via a common perceived stimulus, rather than by direct, individual inputs (Fig. 3.2). Each *on* individual performs a task and by doing so decrements the stimulus level associated with that task by one stimulus unit (negative feedback). When an individual switches *off*, she stops performing the task, and the stimulus level is incremented one stimulus unit. An example of such a negative feedback system in a honeybee colony may be the number of empty food storage cells that stimulates individuals to forage. The perceived, residual stimulus levels change as a consequence of negative feedback. Storage cells fill up as more individuals forage and, with a constant food consumption rate, the

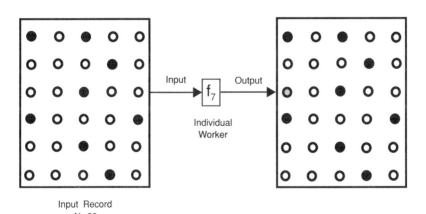

Input Record
N=30

Figure 3.2. How a switching network can be transformed to represent a network of honeybees sharing information through cues provided by the amount of food, say, pollen, stored in a comb. The left box represents a section of comb with 30 cells. The 22 open circles are empty, while the eight closed circles are full. An individual worker assesses the number of empty cells and compares it with its threshold function, f_7. Because the number of empty cells exceeds the threshold value of seven, the worker forages and fills the stippled cell represented in the right-hand box. The record now changes as a consequence of the behavior of the worker.

number of empty or full cells provides a record of foraging activities of all foragers.

Simulation Events

Assume that all individuals are initially *off* and the stimulus level, S, is set equal to N. S is a constant value that is decremented by the activities of individuals that are *on*. The order of events for the simulation is as follows:

1. All individuals of the network (model insects) are assigned a threshold of response f_i randomly drawn from a discrete uniform distribution represented by the set of threshold functions $F' = \{f_i, f_j, \ldots f_n\}$.

2. Individuals are selected using one of the three methods mentioned above: simultaneous, sequential, or random.

3. Each sampled individual is checked to determine whether it is currently *on*. If it is *on*, it is turned *off* and the residual stimulus is incremented one unit. (The residual is the stimulus level minus the number of individuals currently *on*, because we assume that each individual decrements the stimulus one unit while she performs the task. We assume that an individual is not performing a given task while it is collecting and assessing stimulus information.) If the individual is currently *off*, the residual stimulus is not changed.

4. The threshold value for that individual (or each individual in the case of the simultaneous model) is then compared with the current residual stimulus level. If the stimulus level exceeds the randomly assigned threshold value, then the individual is turned *on* and the residual stimulus level is decremented one unit. The individual is then recorded as being *on* or *off* for that sampling event.

Case 1: Variable Data Sampling,
$N = 100, K = N, F' = \{f_1, f_2, \ldots, f_N\}, S = N$

Simultaneous Sampling. We investigated dynamical behavior of model systems over 100 simultaneous sampling events (Fig. 3.3). Simultaneous sampling leads to highly ordered dynamical behavior with great instability. At time t, all N individuals are *off* and the residual stimulus level exceeds the thresholds of all individuals except those that have thresholds of 100 (individuals randomly assigned threshold values of 100 are never *on*). All individuals turn *on* simultaneously and drive the stimulus level toward zero at time $t + 1$. Now the stimulus level is below all thresholds, so individuals turn *off*, subsequently driving the stimulus level back up to 100 at $t + 2$. The system oscillates between nearly all individuals *on* and all individuals *off*.

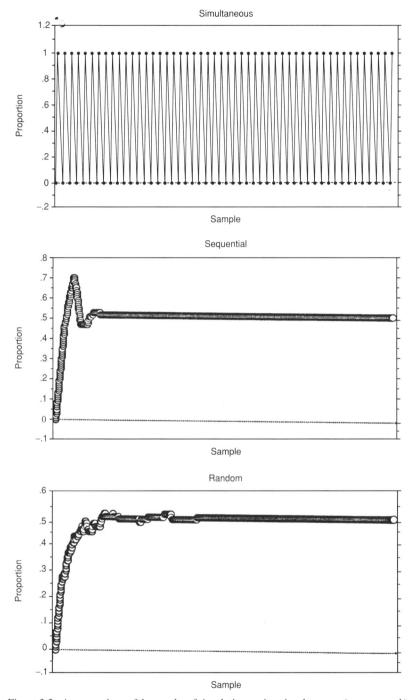

Figure 3.3. A comparison of the results of simulations using simultaneous (*upper graph*), sequential (*middle graph*), and random (*lower graph*) data access sampling of individual network elements. The proportion of individuals that are *on* is shown for each of 100 simultaneous sampling events and for each of 2,000 for the sequential and random sampling models.

Sequential Sampling. For sequential sampling, individuals are assigned numbers from 1 to N (Fig. 3.3). Each individual is sampled in turn, and the order of events described above are followed. A total of 2,000 samples were taken, 20 for each individual. Extremely ordered stable behavior emerges with this system. An initial negative feedback phase occurs that draws the residual stimulus level down near an equilibrium. This occurs as a consequence of individuals turning *on*, so an increase in the proportion of the individuals *on* is observed. However, the system overshoots the equilibrium point and oscillates around it until it hits a steady state attractor. At this time all individuals are frozen either *on* or *off*, showing an extreme division of labor and specialization. In a real insect society, *off* individuals may engage in different tasks.

Random Sampling. It is difficult to imagine how a group of insects could coordinate turnstile-type sequential data access. It seems more likely that some kind of random sampling by individuals will occur by default. For this model system we assume that only one individual samples and changes the residual database stimulus at a time. In this case, connectivity of the network can be asymmetrical: Those that are sampled more frequently, due to chance, have more inputs than those that are sampled less often. Results of 2,000 sample events, an expected value of 20 per individual, are similar to the sequential sampling model. Again, there is an initial negative feedback phase followed by a search for an attractor; however, the search is random rather than oscillatory (Fig. 3.3). It seems that search time to find the attractor may be longer for the random model than the sequential model, but, overall, the system stays closer to equilibrium while it searches.

The sequential sampling model always shows the same system behavior for a given ordered set of individual elements because the sampling order is always fixed. However, the random model behaves differently each time it is run but ultimately ends up at the same steady state value with respect to the proportion of individuals turned *on* (density). Steady state densities in these highly connected systems represent average, typical properties of the networks, the modes of response thresholds among elements. However, the *activities* (how many samples in which an element is *on* or *off*) of individual elements are not closely correlated with their response thresholds because they become fixed *on* or *off*.

Case 2: Variable Sampling, $N = 1,000, F' = \{f_1, f_2, \ldots, f_N\}, S = N$

The dynamical behavior of sequential and random sampling models with $N = 1,000$ was similar to the case of $N = 100$. The sequential sampling systems conduct oscillating searches for attractors, while the random sampling systems search stochastically.

Case 3: Variable Threshold Distributions, Random Sampling, $N = 100, S = N$

Threshold distributions used for random assignment of individual elements can vary with respect to their mean values and their variances.

Variable Mean Thresholds. Dynamical behavior of networks with thresholds drawn randomly from discrete uniform distributions with ranges 1–50 and 51–100 showed the characteristic negative feedback phase followed by random searching near an equilibrium point, then each locked into a steady state attractor. Systems with higher mean thresholds, however, ended up in states with lower densities: In other words, fewer individuals worked.

Variable Threshold Variances. Changing the variance in thresholds around a given mean affects both the rate at which the system locates a steady state attractor and the stability of the system to external perturbation: homeostasis. The time (number of samples) that it takes for a network to arrive at a steady state was determined for discrete, uniform distributions of threshold functions of ranges f_i to f_n and for two-point distributions of f_i and f_n. Thresholds were drawn randomly from these distributions and assigned to each of the 100 elements of the model networks. The interval beween f_i and f_n varied around a constant mean value of 50.5. After each sample event, the simulation checked to see whether the density (proportion of individuals *on*) had changed since the last sample. It was determined that the system had located a steady state attractor after all individuals had been sampled at least once without effecting any change in density. The sample that resulted in the last change in density was considered to be the one that resulted in the system entering a steady state. We ran simulations of 100 random networks for each of 11 intervals for each distribution and determined the average number of samples required to achieve steady state conditions.

Systems with no variance in thresholds locate their steady state attractors quickly and avoid most of the search phase. This is the case with the least number of samples required ($X = 69.3$). However, as soon as variance in thresholds is introduced among elements, the time to locate an attractor increases dramatically (Fig. 3.4). In fact, the systems with the least nonzero variance take the longest time to locate an attractor. Increasing the variance then leads to a decrease in attractor time. Two-point networks locate their attractors faster than networks with elements drawn from uniform distributions with the same range of values.

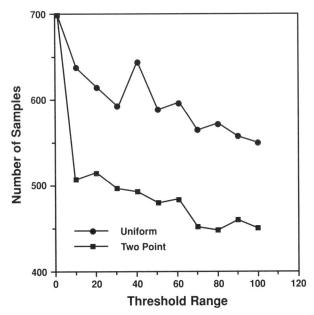

Figure 3.4. The average number of samples, based on 100 simulations each, required for networks to achieve steady states for uniform and two-point threshold distributions with different ranges. The mean threshold for all simulations was 50.5. The case for threshold range = 1 was common for both sets of distributions. The case of threshold range = 0 had a mean value of 69.3 and is not represented.

Case 4: Variable Stimulus Level, Variable Threshold Distribution, Random Sampling, S = N

The variance in thresholds among elements of our model systems profoundly affects dynamical behavior resulting from external perturbations to the stimulus level. Networks of $N = 100$ with thresholds drawn at random from a discrete uniform distribution with range 1–100 were compared with networks where all elements had fixed thresholds of 50.5. The initial stimulus, S, was equal to N, then the residual stimulus level was increased 20 units after every 500 random individual samples. The residual stimulus level was measured at each of 2,500 sample events.

Networks with no variance in thresholds show strong homeostatic properties: They return the residual stimulus to predisturbance levels following each increment of stimulus until all individuals are *on* (Fig. 3.5). Networks with variable thresholds don't regulate the stimulus level as well but do regulate the number of individuals turned *on* or *off* in response to each increase or

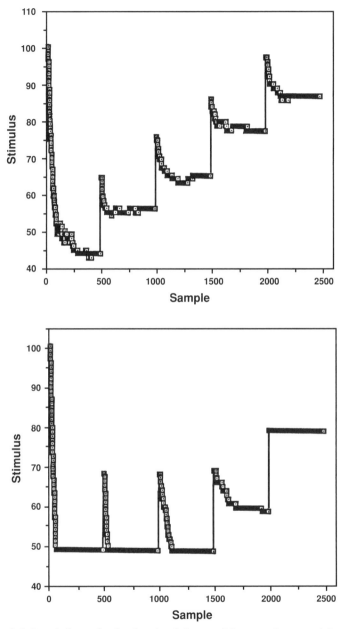

Figure 3.5. Regulation of stimulus level by model networks containing $N = 100$ elements with thresholds drawn at random from a uniform distribution with integer values of 1 to 100 (*upper graph*) and fixed threshold values of 50.5 (*lower graph*).

decrease in stimulus (Fig. 3.5). They buffer individuals from changing stimulus levels and provide a more integrated system response to external changes in conditions.

Case 5: Two Tasks, Two Thresholds, Two Stimuli

The complexity of our model dynamical systems was increased, possibly making them more realistic (but no longer Boolean), by assuming that each individual element could be *off* or *on* in more than one state. This may be equivalent to an individual worker honeybee having the capability of being *on* to task 1 or task 2, or *off*. Individuals that are *off* could be *on* for task 3 or may develop into a different behavioral state (age caste) and become members of different networks. For these models, $N = 100$, and two thresholds were drawn randomly from a discrete uniform distribution with a range 1–100 and assigned to each element. Each threshold was uniquely associated with one of two stimuli that were initially set at 100, and individual elements were sampled at random, as before. Task threshold 1 or 2 for the sampled individual was then drawn at random from a distribution with equal probability and the stimulus level associated with that threshold compared. The individual then responded based on the stimulus-threshold differential. These model dynamical systems again demonstrate the negative feedback and stochastic search phases around an equilibrium value related to the mean of the threshold distribution, but appear to lack easily located steady states (Fig. 3.6). Simulations of 100,000 samples have failed to find steady state attractors; therefore, individuals demonstrated inherently less specialization than did individuals of the single-threshold models.

Sample Both Stimuli. Highly ordered behavior and steady state attractors that result in extreme specialization do emerge from these two-threshold systems (Fig. 3.6) when individual elements check both of their individual thresholds against both residual stimuli and then respond according to the following decision rule: turn *on* to task [i] *if*

stimulus [i] − threshold [i] > stimulus [j] − threshold [j]
and stimulus [i] − threshold [i] > 0.

Thresholds Dependent. Thresholds can be assigned randomly or they can be dependent on each other. We randomly assigned threshold 1, as in the model above, then assigned threshold 2 to individuals on the basis of the following algorithm:

threshold 2 = 101 − threshold 1.

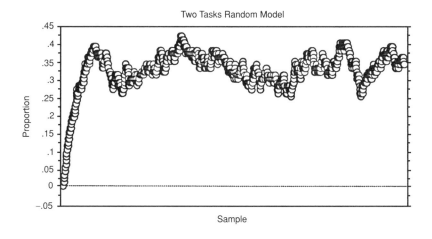

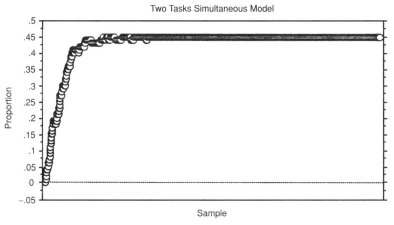

Figure 3.6. Results of simulations of model networks for the case of two thresholds and three possible states for each element (see text). The upper graph is the result when only one stimulus is sampled at a time, while the lower graph shows the dynamical behavior when both stimuli are sampled simultaneously.

This resulted in a reciprocal integrative relationship between the thresholds of an individual with all thresholds of all individuals summing to 100. Integrating thresholds resulted in a greater proportion of individuals *on* for either task at the steady state phase (Fig. 3.7).

Discussion

It is apparent that highly ordered dynamical behavior can emerge from these model networks of individual elements. Even in the simplest systems we see

51

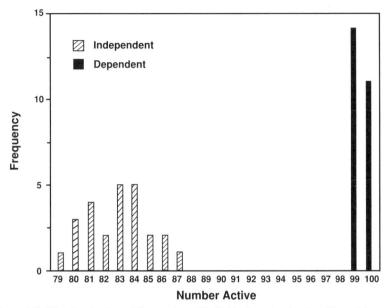

Figure 3.7. The distribution of the number of individuals active (*on* for either of the two tasks) for the two models, independent and dependent assignment of thresholds, from case 5 (see text). The figure shows the frequency distribution for 25 separate simulations.

the kinds of behavior long marveled at in insect societies: (1) **homeostasis** resulting from negative feedback reducing residual stimuli to near an equilibrium determined by the mean of the threshold distribution; (2) **mass action responses** when all elements turn *on* then *off* when simultaneous sampling is employed; (3) system **plasticity** and **resiliency** in response to increases and decreases in external stimuli; (4) **division of labor** and **specialization** demonstrated most dramatically by steady state attractors where some individuals are frozen *on* and others are frozen *off*. Even in cases where we have failed to find steady state attractors, many individual elements with low thresholds remain *on* while those with high thresholds remain *off* and those with intermediate thresholds fluctuate *on* and *off*.

Our models suggest that division of labor with task specialization should emerge among individuals whenever they share a nest (see Page 1997 for examples). Spontaneous division of labor may have complemented the evolution of social behavior from the initial inception of group living but itself may not have been a direct product of natural selection acting on groups because it may be impossible to have groups that do not have a division of labor. This is a result of the stimulus-response-threshold relationship of behavior and because individuals change the levels of stimuli when they respond to them. The number of nestmates, the distribution of response thresholds among nestmates, and the way in which they sample relevant stimuli shape the dynamics of division of labor. Natural selection may continue to modify and fine-tune social organization by operating on the self-organizing features of the complex social system.

Natural Selection Operates on Parameters of Complex Dynamical Systems

In this section we show how natural selection can shape colony organization by acting on heritable variation for queen and worker traits that affect complex system parameters. Natural selection cannot act individually on behavioral components of workers, but instead "sees" the dynamical system properties of colonies. Therefore, natural selection for colony organization may operate on the dynamical system parameters N, K, and $\{F'\}$. Individual colony members may have behavioral and developmental properties that are indirectly selectable and affect these parameters.

Selection on N. For any given distribution of threshold functions, $\{F'\}$, and a residual stimulus level greater than 0, increasing or decreasing N, the number of individuals competent to perform a given task, will increase or decrease the number of individuals *on* or *off* for that task. Colonies with greater numbers of workers are expected to have greater N for a given age caste given a constant rate of behavioral development; therefore, selection for colony size will affect N. Colony size can be affected by both queen fecundity – a trait that may be controlled by workers – and worker length of life, both of which have been shown to have heritable variation (Guzmán-Novoa et al. 1994; Page et al. 1995b).

An increase in N for a specific age caste may also be achieved by altering behavioral development rates so that individuals belong to a given age caste for a longer or shorter period of time (see Robinson 1992 for review). Genetic variability for rates of development associated with age-polyethism have been demonstrated repeatedly (Calderone and Page 1988, 1991, 1996; Giray and Robinson 1994; Guzmán-Novoa et al. 1994; Winston and Katz 1982). N can also be affected by the sensitivity of individuals to environmental stimuli that result in changes in rates of behavioral development into different age castes. Individual workers in colonies that lack foraging-age bees undergo an increased rate of behavioral development until there is a population of workers that forage at a precocious age (Huang and Robinson 1992, 1996). Recent studies have shown genetic variability, hence the potential selectability, for the likelihood that an individual will undergo accelerated behavioral

53

development (Giray and Robinson 1994, 1996; Page et al. 1992; Robinson et al. 1989).

Selection on K. K can be altered by adjusting the cues and/or signals used as stimuli and by changing the data flow operations, for example, the way individuals sample the stimulus environment. Tasks that require mass action – all or nothing – responses should employ simultaneous sampling. Highly volatile chemical signals (pheromones) that are released by individuals as soon as anyone is exposed to those released by other individuals may approximate this data flow structure. This is true for the alarm system of honeybees where a single disturbed individual may release a highly volatile compound that results in the release of pheromone and behavioral responses by all other behaviorally competent individuals that perceive it (Wilson 1971). After individuals release their pheromone and respond, they turn themselves *off* and the colony returns to its original state. With this system, the *on* signals rapidly propagate, but are not instantaneous.

Cues such as the amount of empty or full storage comb in a honeybee nest may provide reliable information about colony needs and the foraging activities of other individuals. The use of these cues should be selected over highly transmissible chemical or mechanical, individual signals when ordered stable behavior is more appropriate. Increasing quantities of brood present in the nest results in more pollen foragers and larger loads of pollen (Eckert et al. 1995), while the presence of more stored pollen results in fewer pollen foragers and smaller pollen loads (Fewell and Winston 1992). The interplay of these stimuli results in the regulation of the foraging population and the pollen intake of the colony (Fewell and Page 1993). Camazine (1993) suggested that nurse bees serve as stimulus intermediaries providing inhibitory cues to foragers that relate to levels of stored pollen and brood; however, there is no direct support for the hypothesis. In any case, the specific mechanism, whether direct perception of stimuli by foragers or indirect correlation of brood and pollen through nurse bee intermediates, does not alter the interpretation of the models.

K can also be adjusted spatially by creating a structural modularity (see Page and Robinson 1991). Individuals located close to each other are more likely to share information – have greater connectivity – than those located in different parts of the nest, thus suggesting a nonuniform assignment of *K* to the *N* nodes. The spatial organization of tasks with respect to where they are performed in the nest may be selectable (Seeley 1982). Calderone and Page (1988, 1991) demonstrated a nonrandom spatial distribution of nestmate workers derived from different genetic sources. Bees that forage on the same plant resources associate with one another in the nest (Lindauer 1961). Oldroyd et al. (1991) suggest that honeybee workers that have the same father (super sisters) are more likely to attend each other's recruitment dances than those of their half-sisters.

Selection on {F'}. Selection on threshold functions may occur by selecting for higher or lower mean thresholds or by selecting to increase or decrease the variance in thresholds within colonies. Changes in mean thresholds result in changes in the likelihood that individuals belonging to specific behavioral castes will perform specific tasks, resulting in concomitant changes in the numbers of individuals engaged in those tasks. Genetic variability for the likelihood that individuals perform specific tasks has been demonstrated in honeybees for pollen and nectar collecting (Calderone and Page 1988, 1991, 1996; Calderone et al. 1989; Dreller et al. 1995; Guzman-Novoa and Gary 1993; Guzman-Novoa et al. 1994; Hunt et al. 1995; Page et al. 1995b; Robinson and Page 1989b; Rothenbuhler and Page 1989), grooming behavior (Frumhoff and Baker 1988), guarding and corpse removal (Robinson and Page 1988, 1995), hygienic behavior (Rothenbuhler and Page 1989), defending the nest (Breed et al. 1990; Guzman-Novoa and Page 1994), caring for queen versus worker larvae (Page et al. 1989; Robinson et al. 1994a), scouting for new nest sites (Robinson and Page 1989a), and other activities performed within the nest (Calderone and Page 1991; Calderone et al. 1989).

Pollen and nectar foraging behavior have been examined in detail. Using the methods of Hellmich et al. (1985), Page and Fondrk (1995) produced artificially selected strains of bees for the amounts of pollen they stored. After three generations, two-way selection produced high and low strains that differed by more than five-fold in quantities of stored pollen. High strain colonies had significantly more pollen foragers and high strain workers individually were much more likely to forage for pollen than were low strain workers, even when raised in a common colony environment (Page et al. 1995b).

Hunt et al. (1995) constructed a genomic linkage map of major quantitative trait loci (QTLs) responsible for observed differences in quantities of stored pollen between the the high and low pollen hoarding strains of Page and Fondrk (1995). They identified two genomic regions, designated *pln1* and *pln2*, that were likely to contain genes that jointly explained about 59 percent of the total observed phenotypic variance. These two genetic regions also affected the likelihood that an individual would forage for pollen. These results suggest that colony level selection affected {F'} by fixing different alleles of *pln1* and *pln2* in the high and low strains.

Page et al. (1998) showed that nectar and pollen foragers from colonies unrelated to the high and low pollen hoarding strains differed in their response thresholds to varying aqueous solutions of sucrose as measured by a simple proboscis extension reflex test. Workers from the high and low pollen hoarding strains varied predictably in their responses to different sucrose concentrations: High strain workers responded like pollen foragers of unselected colonies, while low strain workers responded like nectar foragers. In addition, high and low strain nectar foragers differed in the sugar concentrations of nectar that they found suitable to collect, as predicted by their differences in sucrose response thresholds. They further showed that nectar foragers that inherited from their mother different alleles for QTL *pln2* varied in the sugar concentrations of the nectar they collected in the field, suggesting that *pln2* is directly affecting sucrose response thresholds.

Selection on the variance in thresholds may affect the colony's response to changing environments. Narrow variance may be advantageous when the stimulus itself is being regulated by workers, such as in thermoregulation of the nest. Wide variances may provide a better regulation of work force when it is more important to regulate and integrate the proportion of workers engaged in different tasks rather than regulate the stimulus itself. Large variances also result in more rapid approaches to colony steady state conditions. Selection on the variance may occur by selecting for mechanisms that increase genetic variability within colonies, such as polyandry and polygyny; by selecting heterozygous or homozygous queens (and kings in the case of termites); or by selecting for the number of genetically variable loci associated with specific tasks.

Variance in response thresholds within colonies does not need to be generated by genetic variation. Response thresholds may be changed stochastically among individuals as a consequence of chance individual differences in experience. Behavioral modification through learning may also affect the variance. Individuals may alter their thresholds on the basis of prior experience. For example, an individual's threshold may lower on the basis of performing a specific task, making it more likely he or she will perform that task again the next time the stimulus is encountered. This kind of feedback system would increase the variance in thresholds and serve to functionally wall off those individuals with lower initial thresholds from the other members of the network.

Networks may also be coupled, or *n*-ary as opposed to binary, as we show in our two-threshold model. Individuals may have multiple states with respect to which tasks of a set each performs while belonging to a given age caste. These states may be determined by multiple stimulus-threshold complexes with various decision rules. Natural selection may determine the organization of these complexes forming specific task sets associated with specific developmental stages. Genetic covariance resulting in dependent thresholds of response may increase the overall activity level of individual workers of a given age caste.

For example, dependent thresholds may lead to a greater likelihood that all individuals of the foraging age caste are foraging for something.

For other task sets associated with other age castes, fewer individuals may be needed to perform all the requisite activities, and inactive individuals may undergo behavioral development into the next age caste in the sequence or revert to a previous behavioral state. In this case, independent thresholds of response may serve to provide a pool of relatively inactive reserve bees that have low thresholds of response to most or all task-inducing stimuli and are capable of undergoing behavioral modification through development (see Fergusson and Winston 1985; Robinson and Page 1989a; Robinson et al. 1994b).

Conclusions

The set of models presented here is by no means exhaustive. Additional models can be constructed using different methods of sampling individuals permitting the exchange of information among individuals, coupling tasks, stimuli, and thresholds of response, and using different sets of decision rules. The additional complexity that can be built into the models may make them correspond more closely to real insect societies and may even suggest optimal design features of network systems that may lead to novel empirical investigations. For instance, negative or positive feedback on thresholds of response based on prior experience can easily be incorporated into the models. Individual workers can have different activity levels and subsample the activities of other workers rather than having complete information of the activities of all N subcaste mates. However, it is important to point out that highly ordered system behavior emerged from our model networks that were constructed with minimal lower level order. It would not be surprising to find more high-level order with increasing amounts of order at lower levels.

The models presented in this paper provide transformational processes for building interactive, homeostatic social systems from collections of individuals. These processes are: (1) the stimulus-response-threshold relationship of performing a behavioral act, and (2) the correlation between performing a behavioral act and the stimulus environment. The recognition of self-organization as a potentially important process in the organization of biological systems (Bonabeau et al. 1996; Camazine and Sneyd 1991; Gordon 1996; Page and Mitchell 1991) provides evolutionary biologists with a new approach to understanding the evolution of complex biological design.

4

Evolved Diversity

The array of living organisms that have populated our planet for the last 3.8 billion years constitute neither William James's infant world of a "blooming, buzzing confusion" (1950: 462) nor the perfect, divinely organized "balance of nature" (Egerton 1973). Rather, it is made up of a diversity of organisms showing an assortment of modes of organization solving the problems of survival and reproduction in a variety of ways. Variation is perpetually produced, and competition and natural selection tune that variation to the biotic and abiotic environment in which a population finds itself. Thus, the diversity of life is not random, but neither is it uniform. The sciences that attempt to characterize living things by means of generalizations and laws, common functions, and the operation of a few central mechanisms must face up to the complexity of this domain.

Natural selection, while a major force, is not the only cause of the diverse forms of life. And natural selection can operate at different levels of organization either in concert or in opposition. Attempts to unify the theories of evolution by appeal to a single, privileged unit of selection fail to do justice to hierarchical selection regimes. The first section in this chapter addresses this defining issue in the philosophy of biology and councils pluralism.

Explanations of the evolution of complex traits should not make the mistake of equating any visible trait with an adaptive history. This was the central message of Gould and Lewontin's 1979 "The Spandrels of San Marcos." But, then, how and when can natural selection be invoked in explaining observed traits? The second section in this chapter is concerned with the application of selectionist explanatory schemata beyond morphological characters to behaviors. Here the central theme is the multiple ways in which a behavior can be coupled to its underlying genetic constituent. This type of diversity refutes simple-minded sociobiological accounts of complex behaviors such as rape. The analysis and critique again suggest that the toolbox of scientific explanations in biology must itself be diverse. No simple rule or single mechanism is likely to account for even those traits that appear to us to be similar.

The third section in this chapter addresses a diversity of concepts that is hidden in the use of a single explanatory term, namely, "function." Explanatory uses of "function" are ubiquitous in biology and the social sciences. However, traditional philosophical analyses of this important concept stand at odds with the different explanatory projects in which it is employed. Indeed, analyses in the logical empiricist tradition render talk of function nonexplanatory, which is out of step with scientific usage. More recent accounts of function, including teleological and dispositional interpretations, better cohere with scientific practice. However, I argue that attempts to collapse the different uses of function for explanation into the one, true, best philosophical analysis are bound to fail by ignoring the variety of explanatory work to which this concept is put.

4.1. COMPETING UNITS OF SELECTION? A CASE OF SYMBIOSIS

The units of selection debate is about the nature of the causal processes that constitute evolution by natural selection and the ontological commitments required by those processes.[1] The issue arose in the context of a debate between group selectionists and individual selectionists. In 1962 Wynne-Edwards presented an extensive study of the evolution of "altruistic" behavior in animals. A behavior such as a warning cry was labeled "altruistic" since it appeared to benefit a group of individuals (those hearing the warning) at a cost to the individual actor (the more vulnerable individual who cried out).

Wynne-Edwards suggested that altruistic behavior could evolve only by means of group selection. This mechanism would allow one group to survive rather than another group because the former contained altruistic individuals even though the altruistic behavior was relatively detrimental to its individual carrier. Most generally put, the proposal of group selection posed a deep question about the priority of explanations for Darwinian evolutionary biologists. Which effects of a trait – on the individual organism, on the group in which the organism resides, on the whole species – are significant for evolution?

One response to Wynne-Edwards's claim that group selection is necessary to explain altruistic traits was presented by G. C. Williams (1966). In the spirit of a tradition begun by Fisher (1930), Williams defended a "gene's-eye"

---

[1] This section was originally published as S. D. Mitchell, "Competing Units of Selection? A Case of Symbiosis," *Philosophy of Science* 54 (1987): 351–367.

view of evolution. Rather than appeal to the effects of a trait on an individual organism *or* a group of organisms, Williams argued that evolution by natural selection is best understood as operating on genes. Ten years later, Richard Dawkins further elaborated this perspective in *The Selfish Gene* (1976). What Wynne-Edwards had identified as "altruistic" traits were explained by gene selectionists as having beneficial effects on genes and hence not really altruistic. They proposed that evolutionary change was "for the good of" the gene, rather than either the individual or the group, and hence the gene was dubbed *the* unit of selection.

The question of the appropriate unit of selection has been framed to permit as an answer the identification of a single, privileged, entity: the gene, the individual organism, the group, and so on. Arguments for a given unit have purported to show that reference to the favored entity allows the best perspective from which to understand the process of adaptation, that is, the evolution of traits by natural selection. In this section I argue that with a detailed account of the process of evolution by natural selection on hand, the question of *the* unit of selection turns out to be ill formed. The underlying assumption that there is only one entity essentially involved in the process is unwarranted. A close analysis shows that evolution by natural selection is necessarily a two-step process.[2] Though a variety of entities have been put forward as *the* unit of selection – including the gene, a length of DNA, a chromosome, a genome, an organism, a group, and an entire species (see Gould 1977; Wimsatt 1980) – I consider only two proposals. The first is Richard Dawkins's view that the "gene"[3] is the appropriate unit, and the second is Robert Brandon's defense of the individual organism in response to Dawkins's arguments. I show that although they both explicitly adopt relevantly similar characterizations of the process of evolution by natural selection to the one I defend, they fail to draw the correct implications of that analysis for the unit of selection issue. Thus, my strategy in "solving" the unit of selection problem is to first explicate the causal process in which the unit operates, and then ask what entity or entities must fill the roles described by that process.[4]

---

[2] Mayr (1978) was the first to coin the two-step characterization of evolution by natural selection. His steps are: (1) the production of heritable variation and (2) the testing of variation.

[3] To be fair from the outset, Dawkins does not maintain that a single gene is the relevant entity to isolate. He argues instead that some bit of the string of DNA on a chromosome will replicate as a coherent entity. But in accord with his own choice of simplifying terminology, I use "gene" to refer to the genetic replicator.

[4] Hull's discussion of the "Units of Evolution" (1981) goes far toward clarifying the distinctions between the necessary roles within the process of evolution by natural selection and the metaphysical nature of the entities that can fill those roles.

The Process of Evolution by Natural Selection

Lewontin in 1970 and 1978 gives what has become the standard account of evolution. He says evolution by natural selection is comprised of three individually necessary and jointly sufficient conditions: variation, heritability, and differential reproduction. Mention of adaptation is explicitly excluded from Lewontin's list: "The three principles say nothing . . . about adaptation. In themselves they simply predict change caused by differential reproductive success without making any prediction about the fit of organisms to an ecological niche" (1978: 220). At the same time he admits that "adaptation leads to natural selection" (1978: 221). However, a complete characterization of evolution by natural selection, one that distinguishes that process from other means of evolution, requires an additional principle that specifically invokes adaptation.

It is important to keep in mind that what is at issue in the unit of selection debate is the object of the process of evolution *by natural selection* and not evolution by whatever means. Evolution is the change over time of the frequency of traits of individuals within a population, and it is clear that such change can be brought about by different mechanisms.[5] It can be due to genetic drift or genetic linkage as well as natural selection. The operation of a nonselective mechanism is not relevant to the determination of the unit of selection. The three conditions Lewontin proposes do not rule out evolution by any of these means. An adequate characterization must do so. We must therefore expand the list of Lewontin's conditions to include a principle of adaptation. The added principle requires that the differential transmission of traits across generations is *due to* competition among entities within a shared environment. Relative adaptation describes those interactions of variant individuals with their environment that are responsible for differential reproduction that, given the heritability of the trait in question, will determine which traits are represented with greater or lesser frequency in subsequent generations. As I show, only with the additional condition of differential adaptation will the conditions be sufficient to characterize evolution *by natural selection.*

Consider the following standard example of evolution by natural selection in the wild (see Grant 1963; Kettlewell 1956). The peppered moth, *Biston betularia,* found in England has the habit of resting during the day on the trunks of trees where it is in danger of predation from insect-eating birds. Within this population a speckled gray coloration of wings predominates, though a

---

[5] Here "individual" is to be understood as ontologically neutral with respect to the entity or level of organization. Genes, organisms, groups, and species are all individuals in this sense. See Hull 1981.

melanic form arises as a result of mutation at a single locus. This mutant form has black wings and is more highly visible on lichen-covered trees than the speckled variety and hence more vulnerable to predation in that environment. From the period 1848 to 1898 the melanic form found in populations near Manchester increased in frequency from 1 percent to 99 percent. During the period of evolution the moth's environment changed dramatically. The industrialization of England transformed the originally lichen-covered trees into soot-covered trees. The speckled variety, though virtually invisible on a lichen background, was highly visible on a dark, sooty background. Since the speckled moths were seen more easily under the changed environmental conditions than the black moths, a greater proportion of the speckled variety were eaten by birds before they could reproduce. That was why the black-wing trait came to predominate in that population of moths.

Lewontin's criteria, though met in the standard case described, are insufficient to fully explicate it as an instance of evolution by natural selection. There was variation in wing color between the speckled-gray and the black mutation. This variation was heritable genetically, and furthermore there was differential reproduction between speckled moths and black moths. The satisfaction of the three conditions is necessary for there to have been evolution of black wing color. However, it fails to locate the cause of the change in trait frequency. To show that the evolution was due to natural selection, the additional condition that the interaction of the variant moths and their environment was the cause of differential transmission of traits must also be satisfied. In this episode of the history of the peppered moth, it clearly was. The black-winged moths reproduced more successfully than the speckled moths because the trees on which the moths landed had become dark with soot and so the predatory birds were able to see the speckled moths more easily and ate a greater proportion of them before they could reproduce. In short, black-colored wings were relatively better adapted to the new set of environmental conditions than speckled wings. And this difference in adaptedness was the cause of the evolution of black wing color.

Without specifying the relative adaptedness of the competing traits, the change of trait frequency of wing color in this population of moths might well have been due to a mechanism other than natural selection. It might have been the case, for example, that a homozygous mutant melanic form of *Biston betularia* just happened to be the founders of a small, isolated colony near Manchester. Presuming no emigration from nonmutants and a low rate of further mutation, the population there would be black-winged. Genetic drift can cause the evolution of wing color, but it is not evolution by natural selection.

The distinctive feature that makes the above case an example of evolution *by natural selection* is just what Lewontin's criteria leave out. The fact that the interaction of the competing traits with the environment determined differential reproduction, and hence the change in trait frequency, is a necessary step in the operation of natural selection. Only by appending Lewontin's conditions with a requirement specifying that causal relationship can we secure an adequate account of evolution by natural selection.

In summary, we can see that evolution by natural selection is necessarily a two-step process: (1) The interaction between individuals possessing varying traits in a shared environment, and (2) the differential transmission of the traits to succeeding generations due to that interaction. I argued above for the necessity of the step of environmental interaction. The step of transmission is required, because although differential reproduction may result from the interactions of individuals in an environment, unless the trait responsible for the increased reproductive success is itself transmitted, no change in frequency of the trait (that is, evolution) will result. Imagine that the speckled and black wing color of the moths is determined by two different chemical compositions of the chrysalis. In turn, suppose that the chemical fertilizer used in surrounding fields affects the chrysalis and hence the wing color of the adult moth and that both chemical determinants are present. Suppose, too, that the other environmental conditions are the same as before. In this case there would be variation and there would be differential reproduction based on the interaction of the variants with their environment, but there would not be evolution since the frequency of a given wing color in a future generation would not be determined by the differential mortality of the predecessors. Whether or not ensuing generations were speckled or black would depend rather on the decisions of the farmers in the area, not on the camouflage effect of the trait vis-à-vis the predators.[6] If there is transmission but no interaction, evolution can ensue but it will not be *by* natural selection. If there is interaction but no transmission, differential reproduction will ensue, but it will not lead to evolution. Both steps are required for evolution by natural selection.

Dawkins and Brandon on Evolution by Natural Selection

Dawkins takes seriously the question of what is the unit of selection and has spent considerable time and effort defending the "gene" as the only correct

---

[6] This example is fictitious but not far-fetched. Developmental variation does occur. Arrowleaf plants vary dramatically depending on whether the seed grows on land, partly submerged in water, or completely underwater. See Ricklefs 1973: 216.

answer. In taking this position Dawkins does not by any means present a naive view of the process of evolution by natural selection that ignores the role of individual organisms, groups, or species. Yet, even armed with a characterization of the process similar to the one defended in the last section, Dawkins virulently defends the claim that the status of unit be conferred onto the "gene." In doing so, Dawkins insists that although it is in some sense an issue of "conceptual" preference, it has repercussions on the kind of evolutionary explanations that are produced and accepted.

To understand why Dawkins adopts the gene's-eye view, a look at the way in which he describes evolution by natural selection is necessary. Part of the difficulty in sorting out the points of conflict between Dawkins and Brandon might be due to a lack of shared vocabulary. To bring into sharper focus the differences between the two positions, I attempt to reformulate their views in consistent terms. For example, at different times proponents on both sides of this debate use the terms "evolution" and "selection," *simpliciter,* as well as "evolution by natural selection." It is evident from the argument in the last section and by arguments below that the relevant process to investigate is referred to by the latter.

Dawkins labels the two parts of this process "replicator survival" and "vehicle selection." A replicator is defined as any entity "of which copies are made" (Dawkins 1982b: 46). What is of interest in evolution, however, is a subclass of replicators, namely, "active, germ-line replicators." A replicator is active if it "has some causal influence on its own probability of being propagated" (e.g., genetic material with a phenotypic expression; ibid.: 47). A replicator is germ-line if it is "the potential ancestor of an indefinitely long line of descendant replicators" (e.g., DNA in a germ cell; ibid.: 46). "Replicator survival" is analogous to the second step of evolution by natural selection discussed above, namely, differential transmission. The genetic material that determines melanism in *Biston betularia* would qualify as an active, germ-line replicator. It is both transmitted from a moth to its offspring and phenotypically expressed by black wing color in an environment where that makes a difference. Hence, it influences the probability of its own transmission. However, as we saw, differential transmission alone is insufficient to isolate the cause of evolution.

Dawkins recognizes another part of the process of evolution by natural selection, "vehicle selection."[7] Vehicle selection is defined as "the

---

[7] One can certainly object to the connotation carried by Dawkins's choice of terms. Calling an entity a "vehicle" already presupposes the gene's-eye view of the process, since by definition a vehicle is primarily thought of as a tool of another entity rather than the main actor in a causal scenario.

differential success of vehicles propagating the replicators that ride inside them" (ibid.: 51). Dawkins says little about *how* vehicles propagate replicators or in what sense they can be more or less successful. He tells us nothing of this process outside of its effects on replicator survival. He does, however, believe that vehicles can come in all shapes and sizes. Although the most common vehicle is the individual organism, Dawkins does not want his conceptual schema to exclude from the outset the possibility of a vehicle being a group or a species. (Dawkins notes that empirical matters will need to be considered to decide between them; cf. 1982a.) Vehicle selection can plausibly be taken to stand in for the interaction of variants with the environment. To accept this interpretation is not contrary to Dawkins's views, although it requires some further specification of his account of vehicle selection.

Dawkins discusses variation and competition as requirements of evolution by natural selection, but he is vague about the nature of the competition. He allows, it seems, for the competition requirement to be met by competition between replicators. This, however, would not exclude from the domain of evolution by natural selection the results of the process of gene linkage. The genetic determinants for black wing color might have been linked to the genetic determinants for red eye color. If red eye color was the phenotypic effect responsible for greater reproductive success, black wing color would increase in frequency in the population at the same time. In such a case, a replicator is propagated with greater frequency in a population not because of the interaction of *its* phenotypic effects in the environment, but rather because of its chromosomal proximity to a replicator that does increase in frequency as a result of interaction. The so-called hitchhiker replicator can even be detrimental to the reproductive success of its individual vehicle and still increase in representation because of the gene linkage. Yet, in one sense, the successful hitchhiker replicator is winning a competition among replicators. This type of competition clearly fails to pick out evolution by natural selection. Although it stipulates that a replicator wins a competition, or is relatively more successful than another, it fails to tie that success directly to environmental interaction.[8] Interaction

---

[8] Competition could be interpreted as being competition for location on a chromosome. In that case the gene is both replicator and interactor. The phenotypic effects of the variant replicators must then be specified for that environment, and a causal process explaining how one replicator comes to be present in that site because of those phenotypic properties must be provided. Although it may be plausible to give this type of interpretation for the case of gene linkage, problems arise that even Dawkins finds troublesome in the case of "selfish genes" whose only "function" seems to be as space fillers between other genes. See 1982a: chapter 9.

between variant traits (or in Dawkins's terms, variant vehicles) within an environment is an essential ingredient. By amending Dawkins so that competition is required to occur at the vehicle level (whatever entity turns out to be the vehicle), we can ascribe to Dawkins a conception of evolution by natural selection relevantly similar to the two-step process of interaction and transmission.

Brandon has been instrumental in clarifying the structure of evolutionary theory and explicitly advocates the two-step view of the process of evolution by natural selection (see Brandon 1981). He adds to the Lewontin conditions of variation, heritability, and differential reproduction what he calls "the principle of natural selection." In effect, this principle stipulates that an individual relatively better adapted in a given environment will probably have greater reproductive success than a less well adapted individual (Brandon 1982: 428). Brandon acknowledges the necessity of both the interaction step couched in terms of relative adaptedness and the transmission step of the process. He agrees that both must occur if natural selection is the mechanism responsible for the evolution of a particular trait.

Given the necessity of both transmission and interaction for a process to qualify as evolution by natural selection, an ambiguity arises concerning the *unit* of selection. Is the unit of selection the entity involved in transmission or the entity involved in interaction?

What's in a Unit?

The two-step characterization of evolution by natural selection leads to the view that the original question was ill formed. There is no one unit or entity whose behavior alone defines the process. There are two essential steps – transmission and interaction – and hence two roles to be filled. A single entity may fill both roles, but that is neither required nor made probable by the theory. By interpreting the question of the unit of selection as asking which entities participate in the process of evolution by natural selection, a variety of responses becomes available. A gene, a larger genetic composition, or an entire genome may operate under certain conditions in transmission as an intact unit. An individual organism, a kin group, an unrelated population, a species may operate under certain conditions in interaction with the environment as an intact unit. The unit of selection controversy arose in the context of questioning the form and operation of the causal process of evolution by natural selection. If the analysis given in this paper is correct, a simple univocal answer to the question is inappropriate. Hull puts this point succinctly:

> ... the phrase "unit of selection" is inherently ambiguous. Sometimes it means those entities which differentially replicate themselves, sometimes those which interact with their environments in ways which are responsible for this replication being differential. (Hull 1981: 26)

Dawkins and Brandon both agree:

> The confusion over "units of selection" has arisen because we have failed to distinguish between two distinct meanings of the phrase. In one sense of the term unit, the unit that actually survives or fails to survive, no one could seriously claim that either an individual organism or a group of organisms was a unit of selection: in this sense, the unit has to be a replicator, which will normally be a small fragment of genome. In the other sense of unity, the "vehicle," either an individual organism or a group could be a serious contender for the title "unit of selection." (Dawkins 1982b: 60)

Brandon says:

> ... there is not a unitary unit-of-selection question; the phrase unit-of-selection is ambiguous. One sense has to do with interactors, the other with replicators. (1985a: 94)

A puzzle remains. Whereas both Dawkins and Brandon recognize the ambiguity of the unit ascription, they nevertheless present competing answers to the question of the proper unit of selection. Dawkins repeatedly defends the genetic replicator as the ultimate unit of selection, and Brandon, arguing specifically against Dawkins, claims that "interactors are the ultimate units of selection" (Brandon 1985a: 86). What then, can be meant by "unit of selection"? In general, a "unit" is an individual that participates as a coherent, intact whole in some causal process. Individuation is clearly a theory-relative procedure. What will count as an individual unit will depend on the formulation of the causal process of interest. Since evolution by natural selection necessarily involves two causal steps, interaction and transmission, identification of *the* unit is ambiguous. Hull has argued convincingly that whatever specific entities fill either role in evolution by natural selection, they must be individuals in a metaphysical sense. But genes, gene complexes, genomes, cells, individual organisms, groups, and species all meet this minimal requirement (Hull 1981). To eliminate the ambiguity, Dawkins appeals to "adaptation" as the arbiter of unit status.

The Unit of Adaptation

Dawkins advocates the view that

> [a]ctive, germ-line replicators . . . are units of selection in the following sense. When we say that an adaptation is "for the good of" something, what is the something? Is it the species, the group, the individual, or what? I am suggesting that the appropriate "something," the "unit of selection" in that sense, is the active germ-line replicator. (Dawkins 1982b: 47)

Hunting for the unit of selection is transformed into a search for the entity or entities that benefit from possession of an adaptation.

There is no dispute that adaptations are phenotypic traits, that is, traits of interactors such as black wing color or antler size or warning cry behavior.[9] Not every interactor trait counts as an adaptation. Only such traits that have evolved by means of natural selection are adaptations. Thus, to identify a trait as an adaptation is to make a claim about its causal history. In fact, it is the appeal to a special type of causal history that makes sense of the "for the good of" language that Dawkins introduces. "Adaptation" is a tele-ological ascription (see Brandon 1981; Wright 1976) and hence can be ex-plained by appeal to a consequence of the adaptive trait. If the conditions specifying the process of evolution by natural selection are met, the con-sequence of having a trait will be the reason that trait appears with greater frequency in subsequent generations. With that selection background em-pirically confirmed, adaptation questions about why peppered moths in the population near Manchester have black rather than speckled wings can cor-rectly be answered by citing the camouflage effect of black wing color in that environment. Thus, an adaptation is a trait of the entity whose interac-tion with the environment determined the differential transmission of that very trait.

Having diagnosed the problem of the unit of selection to be the prob-lem of the unit of adaptation benefit, Dawkins then defines benefit in terms

---

[9] The genotype/phenotype distinction is traditionally drawn between the genetic properties and the "visible" properties of an individual organism. With the more general analysis of the pro-cess of evolution by natural selection on hand, this distinction can be profitably recast. What is of interest are the differences between properties of transmitters and properties of interactors. Since a variety of entities can be classified as transmitters or as interactors, the individual or-ganism point of view that echoes in the genotype/phenotype language may be as misleading as the gene's-eye point of view pervading the replicator/vehicle language. Dawkins, in particular, attempts to extend the range of phenotypic traits beyond those directly attached to an organism's body.

of longevity or survival:

> Replicators that tend to make the successive bodies they inhabit good at avoiding predators, attracting females, etc., tend to have long half-lives as a consequence. (1982b: 49)

> Germ-line replicators . . . are units that actually survive or fail to survive, the difference constituting [evolution by] natural selection. Active replicators have some effect on the world that influences their chances of surviving. It is the effects on the world of successful germ-line replicators that we see as adaptations. Fragments of DNA qualify as active germ-line replicators. (1982a: 95)

In point of fact, only when the unit is understood to be a replicator type rather than replicator token can we talk of its survival. Dawkins implies that a replicator survives or not by means of the presence of "copies" of it. But it is not the original molecules that survive, only other tokens of that type.

To be precise, we must say that the type of DNA that codes for the trait in question becomes more prevalent, or tokens of that type of DNA increase in frequency relative to tokens of other types. Even though Dawkins is quick to point out that an individual organism does not exist beyond its lifetime and hence could not be around in the flesh to reap the benefits of adaptation, neither do physical bits of DNA. When the type/token distinction is acknowledged, appeal to longevity as the benefit accrued by adaptation fails to cut between transmittors and interactors. One could as easily say that a type of individual, the black-winged type, for example, "survives longer" since it becomes more prevalent in a population through time, or tokens of that type of individual become frequent relative to tokens of another type. What replicates directly is a physical entity, a composite of DNA molecules. What interacts directly with its environment most commonly is an individual organism. Yet what survives through time is a type of genetic material or a type of individual organism.

Dawkins argues that complexes of genetic material can qualify as replicators because they have sufficient fidelity through transmission. If there is a change in the gene complex, that change will be passed on. If there is a change in an organism, or a group, or a species, the change will be transmitted *only* *if* it is mirrored by a difference in the genetic associate. Thus he says, "There is a causal arrow going from gene to bird, but none in the reverse direction" (Dawkins 1982a: 98). It is certainly the case that the entity that is most directly transmitted is the replicator, and its structure is passed on without much deviation. One might call this fidelity through transmission. Yet on the basis of this type of fidelity that individual organisms (or groups or species) clearly lack, Dawkins argues that "[i]t is therefore better not to speak of adaptations as being for the good of the organism (or group or species)" (1982a: 99). This does not follow.

Whereas genes operate as coherent entities in transmission, organisms or groups or some other entity may operate as integral wholes in interaction with the environment. If fidelity through transmission is the criterion for recipient of adaptation advantage, by stipulation the replicator will be the only entity qualified for such benefit. But as all the arguments so far have shown, transmission is not the only causal process required. Whatever fidelity in transmission a replicator has, to fulfill the destiny of survival, appropriate events have to occur to its corresponding interactor. Coherence in interaction with the environment is just as clearly characteristic of interactors.

Dawkins's arguments for why the gene is the appropriate unit of selection collapse into arguments for why the gene is the appropriate unit of replication. Nothing is added by his appeal to adaptation as adjudicator of the true unit of selection. All the requirements stipulated for a recipient of adaptation benefit – longevity and coherence – are ascribable equally to transmitters and interactors.

Brandon, in his paper "Adaptation Explanations: Are Adaptations for the Good of Replicators or Interactors?" (1985a), takes Dawkins to task for his defense of the gene as the unit of selection. He argues that even if one accepts the reformulation of the problem in terms of adaptation benefit, the individual organism will reign triumphant in the role of unit of selection. Brandon bases his claim on explanatory priority. Differential transmission of replicators, he claims, is explained by differential reproduction of interactors, and not vice versa. To make his case for the interactor, Brandon uses the "screening-off" analysis developed by Salmon and Reichenbach (see Salmon 1971). Screening off describes differences between two properties in affecting the probability of some event. A screens off B from E if the probability of E, given A and B, is equal to the probability of E given A, but is not equal to the probability of E given B:

$$Pr(E, A \, and \, B) = Pr(E, A) \neq Pr(E, B)$$

If A screens off B from E, then B is statistically irrelevant to E. Whenever A is already present, the addition of B fails to change the probability of the effect. The converse is not true. When B alone is available, the probability that the effect will occur will be increased if A is present as well.

Brandon uses this type of analysis to compare the statistical relevance of traits of interactors and traits of transmitters. He argues that interactor traits screen off transmitter traits from the effect of reproductive success.

Thus, they are more important referents in explanations of differential reproduction. Taking genotypes and phenotypes as the paradigmatic examples of
transmitters and interactors, Brandon shows that an individual organism will
reproduce more or fewer offspring depending on its phenotypic properties,
irrespective of its genotypic properties.

Take, for example, the case of a facultative trait where the same genetic
composition has two different phenotypic expressions. Suppose that one
of the expressions is better adapted to the environment than the other and
hence individual organisms possessing that expression will reproduce more
successfully because of it. In such a case there are identical genotypes and
different phenotypes. There is differential reproduction, and so it must be due
to the differences of traits at the level of interactor since there is no difference
at the level of transmitter. The probability of successful reproduction of
offspring of a given individual organism (the effect, E) will depend entirely
on which phenotypic expression is present (A). Given one knows which
expression is present, addition of information about the genotype responsible
(B) will not change that probability. Indeed, knowledge of only the genotype
present (B) will lower the probability of obtaining the effect (E) since both
phenotypic expressions are associated with it. In this case, phenotypic traits
screen off genotypic traits.

Consider the converse situation in which two different genotypes give
rise to the same phenotypic expression. In such a case, the probability of
successful reproduction of a given individual organism (the effect, E) will be
the same whichever genotypic trait is present. Again, phenotypic traits screen
off genotypic traits since here the difference in genotype has no effect on the
probability of reproductive success.

There is nothing wrong in Brandon's use of the screening-off argument.
However, it is restricted to the statistical relationships relevant to the reproduction of offspring. Dawkins could construct a parallel, opposing argument
using the same tactics. One could argue that the active, germ-line replicator
screens off the vehicle from differential transmission. To see this clearly, recall
the case of gene linkage. Here certain replicators are transmitted at a greater
rate than others, irrespective of their phenotypic expression. The probability
of transmission depends only on replicator properties (that is, position on the
chromosome), and phenotypic expression is statistically irrelevant. Knowing
what phenotypic expression corresponds to the hitchhiker gene will make no
difference to the probability of its transmission, whereas knowing its position
on the chromosome (given the information required to classify it as a hitchhiker in the first place) will make a difference. Replicators screen off vehicles
for differential transmission.

One might well object that the case contrived in support of Dawkins's position, though qualifying as an instance of evolution, is not by natural selection and hence should not carry much weight in the unit controversy. But again, there is a parallel retort to Brandon's screening-off argument. In the case of the facultative gene that had two different expressions, there would be differential reproduction based on differential interaction, but there would be no consequential evolution because there would be no directly proportional corresponding differential genotypic replication. In this type of case there is no evolution. It is only when differential interaction determines differential transmission that we have evolution by natural selection. Both steps must be present and they must be linked. The screening-off argument works for Brandon only when there is no evolution because there is no differential transmission. The screening-off argument works for Dawkins only when there is evolution but it is not due to natural selection.

I have shown that Dawkins and Brandon both accept the two-step analysis of evolution by natural selection, yet they fail to apply that analysis at crucial times in arguing for the appropriate unit of selection. What is going on is that Dawkins and Brandon are addressing different questions when they ask which entities benefit from adaptations. While both recognize the same two-step process as defining evolution by natural selection, they shift their causal focus to only one or the other of those steps in searching for the unit of selection. What effect of adaptation is relevant to determining who benefits? The process as they both see it is illustrated in Figure 4.1. Brandon asks what effect adaptation has on the differential reproduction of organisms, while Dawkins asks what effect adaptation has on the differential transmission of genetic replicators. The point is that when we identify a trait as being an adaptation, both effects are necessary. Neither a case of differential reproduction without the corresponding differential transmission nor a case of differential transmission without the corresponding differential interaction will be relevant, as neither alone results in adaptation.

Brandon correctly answers the question he poses, namely, which entity, the transmitter or the interactor, most immediately benefits in the form of differential reproduction by the possession of an adaptation? Furthermore, differential reproduction at a given time will indeed determine the differential transmission of genetic replicators. Yet Dawkins's account correctly answers the question of which entities, because of their fidelity in transmission, can be thought to survive in the form of copies as a result of possession of an adaptation. Adaptation is "for the good of" interactors if the good is to be measured by the number of offspring produced. Adaptation is "for the good of" transmitters if the good is to be measured by the frequency of entities with

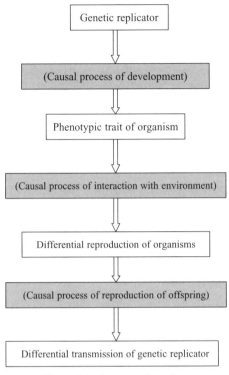

Figure 4.1. Steps to adaptation.

virtually similar structure transmitted to succeeding generations. Screening-off arguments fail to make the case for one or the other.

Brandon suggests another rationale for adopting the interactor as the unit of selection. By so doing, selection at different levels of organization can be distinguished by the entity identified as the unit in any given case. If it is an individual organism's trait that is responsible for differential reproduction, we would call it individual selection, whereas if it is a trait of a group of organisms that determines differential reproduction of the group, we would call it group selection. Such distinctions do need to be drawn, but they can be drawn adequately by identifying different levels of interaction. To ascribe the status of "unit" to the entity of interaction in order to label this difference continues to blur the necessity of the relationship between interaction and transmission.[10]

---

[10] Again, a Dawkinsonian counterargument could be launched. Why not reserve the title of "unit" for the one entity that is constant throughout? Whether or not interaction takes place at the chromosomal, individual, or group level, for the process to be evolution by natural selection, the one thing that always must be affected is a corresponding change in an entity of transmission.

Entities are required to fill both roles. Debates on *the* unit of selection shift attention away from the dual nature of the process of evolution by natural selection. By instead committing oneself wholeheartedly to accepting the necessity and importance of both processes of transmission and interaction, not only can the distinctions elicited by each of the steps be drawn but the symbiotic relationship between the steps can be discerned.

Identifying Adaptive Traits

When Dawkins champions the active, germ-line replicator as the unit of selection, his arguments make interaction look negligible:

> Intergroup selection and interindividual selection . . . are different proximal processes whose claim to biological importance is judged on the extent to which they can be shown to correlate with what really matters – interallele selection. I believe it is often superfluous, and sometimes actually misleading, to discuss natural selection at these higher levels. (Dawkins 1978: 127)

This is certainly too strong. Some account of interaction is already embedded in his definition of the relevant entity. An "active" germ-line replicator must have some effect on the probability of its own replication. If the effect is constitutive of the process of evolution by natural selection, then to be active is to have a corresponding interactor whose relationship with the environment "actively" engineers the probability of replication. Without an account of this aspect of evolution by natural selection, an "active" replicator could not even be identified.

Similarly, for a phenotypic trait to be identified as an adaptation, more than just its effect on the reproductive success at the interactor level must be known. An adaptation is not just any property that makes an interactor better at reproduction. It is a property whose predominance in a population is explained by that ability. It must be transmitted with greater frequency as a result of the interaction effects or it will fail to have the appropriate causal history that qualifies it to be called an adaptation. Dark wing color allows moths to be better, on average, at escaping predators. This is an adaptation because the interaction with the environment is one that affects the propagation of the trait. It clearly does so by means of differential mortality that has an obvious, though not trivial, relationship on differential transmission. We pick out a trait as a potential adaptation because its effects can influence transmission in an appropriate way. Whether it does (whether black wing color in this specific case is adaptive) will require a detailed ecological study of the relationship between interactor and environment *and* the relationship between expressed trait and transmitted information. Even so, the identification of a trait as a candidate for adaptation ascription presupposes some account of the process of transmission.

To see this more clearly, consider a nonbiological case. If individuals who die in battle with an enemy are revered by the survivors and their characteristics thus emulated, one would predict a higher frequency of the characteristics of the dead warriors to occur in the next generation. Mortality in this case does not negatively affect transmission. Of course, a different adaptation story must be told, but it will still require steps of interaction and transmission. Selection by nature requires selection criteria. The interactor/environment relationship has a myriad of effects.

Only those relevant to transmission will participate in the process of evolution by natural selection. In the case of the peppered moth, escape from predation, given our knowledge of the transmission process, is obviously crucial to transmission because differential mortality prior to reproduction directly affects transmission. In other situations, effects like larger clutch size or exploitation of more food resources may not be as direct. They, too, will be relevant only if they have some effect on transmission. To determine whether a trait is adaptive, one must show both that the trait has the appropriate effects in competition with the environment and that having those effects has consequences for transmission.

The analysis of evolution by natural selection as a two-step process allows the question of *the* unit of selection to be solved by being dissolved. Entities are required for both steps. Dawkins and Brandon, in their efforts to defend competing units of selection, have contributed substantially to our understanding of the processes of transmission and interaction, respectively. Even so, the search for an ontologically privileged unit threatens to distort the complexity of the process and obscure important relationships between transmission and interaction. In giving up the search for a single unit, evolution by natural selection and its requisite ontological commitments are better understood.

4.2. THE UNITS OF BEHAVIOR IN EVOLUTIONARY EXPLANATIONS

Sociobiology is that branch of evolutionary biology that aims at providing biological explanations of social behavior.[11] Sociobiology invokes no new general theories. Rather, it is characterized by its special domain. Given the

---

[11] This section was originally published as S. D. Mitchell, "Units of Behavior in Evolutionary Explanations," in Dale Jamieson and Marc Bekoff, eds., *Interpretation and Explanation in the Study of Animal Behavior* (Boulder: Westview Press, 1990), 63–83.

assumption that natural selection has been the most significant force operating in evolutionary history, the explanation of the presence of a given behavior in a population is most often couched in terms of its adaptive significance. That is, given recent developments in evolutionary biology, the explanation of a behavior can employ any of a variety of analyses (including game-theoretic models, optimality models, kin selection models, and reciprocal altruism models) to show how a given behavior, in a particular environment, affects the reproductive success of individuals who display that behavior. In this regard, sociobiological investigations of adaptive behaviors require the same evidence as other evolutionary inquiries, including measures of the consequences of trait possession on relative reproductive success, the genetic basis of the behavior, the historically available alternatives, and the level at which selection is operating.

Though detailed evidence for all the parts of a justification of adaptive significance is difficult for any evolutionary explanation, sociobiological explanations are subject to further, domain-specific, complications. While behavior is unquestionably part of an organism's phenotype (or a gene's "extended phenotype") (Dawkins 1982a), I argue that special concerns regarding the target of selection in sociobiological explanations, that is, individuating evolutionarily significant behaviors, are problematic. In this section I consider two such problems: the difficulties of individuating evolutionarily significant behaviors and the collateral problem of recognizing similarity of behaviors across species. After a general discussion of these issues, I turn to some recent studies of the adaptive significance of rape for illustration.

Causal Explanation in Evolutionary Biology

To claim that this or that behavior is an adaptation, rather than an aberration, or present just by chance, is to invoke a particular causal history. Here "adaptation" refers to a result of the historical process of evolution by natural selection. This usage of the term is common, although there are instances where "adaptation" is taken to refer generally to the "fit" of organism to its environment, whatever the causal process that generated it. According to the second sense, the chameleon "adapts" to the change of color of the background by the chemical process leading to a change in skin color, and peppered moths adapt to environmental pollution by changes over time in frequencies of melanism in the population. In this section I follow the first interpretation (Brandon 1985a; Burian 1983; Gould and Vrba 1982; Mitchell 1987b; Williams 1966) and embrace the historical connotation. Identifying a behavior as an adaptation then can be taken to offer an answer to the question,

Why is this behavior, rather than another, present in the population? Correctly identifying a behavior as an adaptation entails that this behavior, rather than some historically available alternatives, has evolved by means of natural selection because of its consequences on reproductive success in a specified environment. Of course, cases other than fixation of a single trait in a population can result, for example, frequency-dependent selection issues in a population maintaining a variety of traits that are adaptive at specific frequencies.

"B_1 is an adaptation" entails that: B_1 is present in a population because, relative to historically available alternatives $B_2, B_3, \ldots, B_n$ in environment E, B_1 yielded, on average, greater net inclusive fitness than $B_2, B_3, \ldots, B_n$.

We can separate the required evidence into three parts:

1. Showing differential reproductive success results from having or not having B_1;
2. Showing the proximate mechanisms of the behavior/environment interactions on reproductive success;
3. Showing differential genetic transmission of B_1 results (and hence leads to differential expression of the phenotype).

Meeting conditions 1 and 3 ensures that B_1 will increase in frequency in the population over time, that is, it will evolve. Condition 2 is required in order to distinguish selection *for* B_1 from mere selection *of* B_1. This extremely useful distinction was drawn by Elliott Sober (1984c) in order to clarify the nature of the causal process generating a particular trait. It allows us to distinguish mere evolution of a trait from evolution by natural selection for the trait. To be an adaptation, the behavior must be a direct result of evolution by natural selection. It must be B_1's relation to environmental conditions that results in relatively higher reproductive success, rather than its being associated with reproductive success by means *of* either an indirect selective process, such as chance (by drift), or an indirect selective process (B_1's being genetically linked to another trait and then increasing in frequency when the linked trait is directly selected for its consequences in that environment). In short, the explanandum behavior must result in the alleged consequence on reproductive success, and that consequence must be directly causally relevant to the presence of that very behavior.

The difficulties in directly justifying an adaptation explanation are legion (Endler 1986). Indirect arguments for adaptive significance are also offered. Two types of indirect argument – comparative analysis and the bypassing of proximate causal mechanisms – are common in extending sociobiological explanations from the nonhuman to the human realm. Comparative analysis or analogical arguments are proposed to allow inference about the evolutionary significance of a trait from evidence gleaned from multiple populations or species. Sometimes the adaptive significance of a trait is inferred from the correlation between repeated instances of it and a specific ecological condition. Given the correlation it is inferred that the trait has evolved as an adaptation in response to that ecological condition. This inference assumes that similar selection pressures produce similar responses. Divergent pressures produce divergent responses. (See Bock 1977 for a detailed account of the assumptions of this type of argument.) On other occasions, direct evidence of the adaptedness of a trait to certain environmental conditions obtained for one species is generalized to other species sharing the same trait and conditions. All such comparative arguments are based on the presumption of similarity of traits, ecological conditions, and selection pressures.

A second type of indirect argument is found almost exclusively in human sociobiology. It allows evidence of the reproductive consequences of alternative behaviors to justify ascription of adaptive significance by presuming that there must be some genetic basis for all phenotypes, and hence whatever developmental sequence or environmental trigger directly causes the phenotype can be ignored. That is, the "ultimate" causes for any trait are based in the genetic substrate, so "ultimately evolutionary explanations can ignore the proximate mechanisms" (Durham 1979; Irons 1979).

Clarifications of the evidence required "ideally" for justifying a claim that a trait is an adaptation have been developed, in part, in the context of distinguishing between different *levels* at which selection may operate. For example, describing the benefits as accruing at the group level, and the differential transmission of the traits by means of differential group propagation, gives grounds for claiming the trait is a group adaptation and hence that selection has operated at the group level. This section is not concerned with questions of the level of selection but rather with the *target* of selection. Which behaviors are candidates for explanation by appeal to the process of evolution by natural selection (at whatever level)? In the case of direct experimental evidence, one may ask what counts as a behavioral "unit" that could be an adaptation. For indirect inference one must consider what counts as the "same" behavior in different species. I consider these two questions in turn.

Individuating Behaviors

The "Adaptationist Program" has been criticized for accepting evolutionary explanations that presume that every observable trait is adaptive and then conjure a story that justifies that assumption. One objection to this strategy concerns the "atomization" of traits. By assuming every conceptually distinct trait is an adaptation, the argument goes, we have made errors in identifying the actual objects of biological processes. Gould and Lewontin (1979) have argued that what appears to us as an individual trait may not always appear so to the forces of evolution. We intuitively begin by suggesting that what is perceptually distinct to us (such as the size or color of an individual) is a trait that has evolved by means of natural selection. But this may fail to explain the presence of the trait, because what we have identified is not an adaptable unit. One example they offer is the change in accepted explanation for the shape of the human chin. While we can designate a portion of the anatomy of the face as "the chin" and can see variety in this feature within a population or over evolutionary time, it does not operate as an integral whole in the process of evolution by natural selection. What we see as "the chin," the process of evolution "sees" as the necessary consequence of two distinct growth fields. Gould and Lewontin claim that the chin is a developmental artifact of evolution operating on other discrete traits and not itself the object of the joint process of selection and evolution. Since adaptations result only when both processes operate on the same object, the chin fails to be an adaptation.

This example challenges the strategy of producing adaptation explanations for ignoring developmental constraints. We cannot tell an adaptation story about the chin, it is claimed, because selection cannot weld together what development has torn asunder. (But see Alcock 1987 and Gould 1987a, b for a disagreement about the significance of this type of claim.) I suggest that adaptation explanations of social behavior might suffer a related hazard. In this context, adaptive status is conferred onto individual behaviors that may, in fact, not be individually transmitted. An isolated behavior, such as rape, may not be the correct subject of an adaptation story if it is an integral part of a complex behavioral strategy or the outcome of a learning process. While isolated behaviors may be shown to have the requisite differential consequences for reproductive success, the genetic transmission of the behaviors may take a more complex route.

The possible relationships between genetic replicators and individual behaviors increases in number and complexity when we consider the role of learning in generating specific human behaviors. For the purposes of this discussion I am not concerned with the complexities occurring in the genome, that is, whether a single allelic pair, multiple alleles closely aligned on the chromosome, or a more complex interplay among disparate sections of the chromosome control phenotypic expression of a given trait. Rather, my concern here is the complexities at the phenotypic level and the path from whatever the relevant replicating structure in the genome that is central.

Thus, "gene" or "genetic replicator" is used in the sense employed by Dawkins (1982a) and Hull (1981). Direct evidence of the genetic basis of many behavioral traits has been difficult to obtain. Given the variety of pathways from genetic replicators to behaviors, evidence of reproductive consequences alone will not be sufficient to endorse claims of adaptedness.

From Genes to Behavior

To evaluate the legitimacy of such explanations, it is thus necessary to explicate the variety of possible causal pathways connecting genetic replicators and social behaviors. If phenotypic variation is the direct object of natural selection, one must understand the underlying relationship between the phenotypic expression and genetic replicators to argue that any such phenotypic trait is, or can be, an adaptation.

One-One Relationship: g *iff* b_1 *and* g' *iff* b_2. If it is plausible to assume that specific behaviors are genetically defined directly with little environmentally induced variability, then, given a history of genetic variation, the presence of a behavior in a population can be unproblematically explained by its effect of maximizing inclusive fitness. Obviously, traits are neither completely genetically determined nor completely environmentally determined. Everything always has genetic and environmental components. The question is, rather, when should we appeal to a genetic component to explain the presence of the trait, and when to an environmental component? If specific behavior b_1 is directly tied to a replicator g, and b_2 is tied to g' in a one-one relationship, then it is clear that the differential consequences of having b_1 relative to b_2 on an organism's reproductive success will cause one, say b_1, to be present in a population via evolution by natural selection. (See Fig. 4.2, pathway I.) In this case, the explanation of a behavior can be identified by its consequence on reproductive success, and hence the adaptation claim is justifiable.

A one-one relationship is found in traits such as wing color in peppered moths. Any trait will have a range of expression depending on the range of environments experienced during development, that is, the norm of reaction of the trait. Phenotypic expressions are the result of both the genetic coding and the environment of development and expression. The same genotype developing in two different environments can have very different corresponding phenotypes. (See Ricklefs's 1973 discussion of the arrowleaf plant for an illustration.) Once the organism is developed, the trait will no longer vary substantially with variation in the environment (see Fig. 4.2, environmental input between g and b_1). This category, however, cannot be applied to social

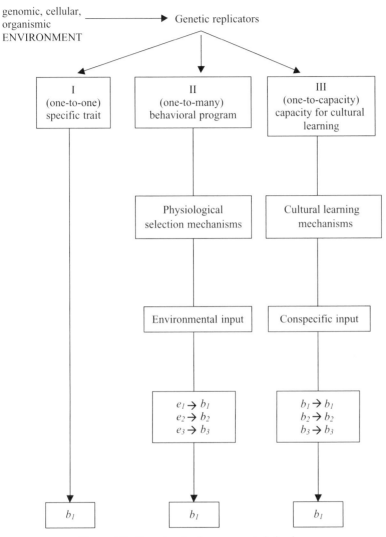

Figure 4.2. Causal paths from genes to behavior.

behaviors. It is in the very nature of what it is to be a social behavior, rather than a fixed trait, that the action involves a relationship with at least one other individual and is in part environmentally induced. The behavior will vary at least with respect to the presence or absence of relevant environmental input after development. This seems a plausible necessary condition even given our lack of detailed information of exact causal pathways from genetic replicator to behavior. "Our ignorance of the pathways from genes to morphology is great. Our ignorance of the pathways from genes to behavior, pathways which surely vary with differences in the achieved morphology, is even greater" (Burian 1981: 54–55).

One-Many Relationship: If g *then* b_1 *or* b_2 *or* b_3 *and if* g′ *then* b_1' *or* b_2' *or* b_3'. Suppose that specific behaviors are not tied directly to specific genetic counterparts, but are rather the result of environmental inputs via a genetically determined proximate mechanism. (See Fig. 4.2, pathway II.) Thus, behaviors are facultative, rather than obligate, or are governed by a "closed program." Here a particular *g* codes for a range of possible behaviors. Which behavior is expressed requires additional environmental input. Such environmental information is then mediated through proximate physiological mechanisms to generate a specific behavioral output:

$$
\begin{aligned}
If\ g &\rightarrow \text{if } e_1 \rightarrow b_1 \\
&\quad\ \text{if } e_2 \rightarrow b_2 \\
&\quad\ \text{if } e_3 \rightarrow b_3 \\
if\ g' &\rightarrow \text{if } e_1 \rightarrow b_1' \\
&\quad\ \text{if } e_2 \rightarrow b_2' \\
&\quad\ \text{if } e_3 \rightarrow b_3'
\end{aligned}
$$

A specific behavior is expressed only by having both the genetic capacity that codes for the environment/behavior pair and having the appropriate environmental input. Thus, behavioral variation within a population may be the result of either the variation in experiential histories of different individuals, that is, differences in *e*, or may be due to similar environmental experiences and differences in the genetic replicator. If there is (or was) genetic variation (one of the prerequisites for evolution by natural selection), then the adaptive significance of a strategy including a set of behaviors can be identified with that set's consequence on reproductive success and contributes to the justification of identifying the complete strategy as adaptive.

Since each individual acquires a behavior via the appropriate gene/environment conditions, the only means of transmission of traits across generations is the genetic pathways through differential replication. Thus, evolution by natural selection is the appropriate causal history. Since that process takes consequences on reproductive success to be causally relevant, those consequences are explanatorily relevant.

The behavioral response in this case is "hard wired" by the genetic program, what Mayr calls a "closed program" in which "the program is contained completely in the fertilized zygote" (Mayr 1974a: 652). Evidence for the genetic component of the behaviors in category II can be obtained from studies involving artificial selection. For example, it was noticed that there are two mating strategies adopted by male field crickets (*Gryllus integer*). A male will call either frequently or infrequently. A breeding program was developed to test for the genetic component to these alternate behaviors (there might clearly be environmental components having to do with the density of crickets in the area). Intense artificial selection effected a change in the frequency of the calling trait very quickly and hence provides evidence of a genetic component. (This case was reported in Trivers 1985: 95–98.)

Environmental Learning plus Cultural Transmission: If g *then [(*$e_1 \rightarrow b_1$*) or (*$e_2 \rightarrow b_2$*) or (*$e_3 \rightarrow b_3$*) or (*$b_1 \rightarrow b_1$*) or (*$b_2 \rightarrow b_2$*) or (*$b_3 \rightarrow b_3$*)].* Consider yet another way in which a specific behavior can be acquired. This is what some have called an open program, cultural learning, or cultural transmission. Here the genome determines a capacity for learning from the environment or conspecifics. One manner in which learning occurs allows an individual to adopt behavior b_1 by imitation of another instance of b_1 expressed by a conspecific. There clearly are genetic constraints even on such an open behavioral program. "An open program is by no means a *tabula rasa*, certain types of information are more easily incorporated than others" (Mayr 1974a: 652). Here the informational input (a subset of general environmental input, namely, that which is specifically from the behavior of another conspecific) is mediated through some proximate learning mechanism that selects a behavior as preferential to others based on some specified selection criteria (see Fig. 4.2, pathway III). The specification of how this mechanism operates is the domain of learning theory. Some proximate selection criteria suggested include avoidance of pain and maximization of "satisfaction."

As outlined above, for a trait to be an adaptation, two processes must occur and be appropriately connected. The first is that interaction of variants in a given environment resulted in differential reproduction. Whether it be through an individual trait bearer's own reproduction or that of genetic relatives, to be an adaptation the trait must have such an effect on reproductive success. The second is that the trait be differentially transmitted, which will occur if the trait is genetically produced and there is differential reproduction. That effect on reproductive success must cause the differential transmission of genetic replicators. And the differential replication of genes must in turn be responsible for the consequent frequencies of the behaviors.

It is obvious that different behaviors can have varying effects on the reproductive success of an individual and his or her genetic relatives. Abstaining from reproduction and not aiding in the survival and reproduction of kin is a behavioral trait that will fare ill on this test compared with having lots of offspring and helping kin. But the celibate hermit's behavior would be maladaptive in the biological sense only if the genetic replicators that get transmitted differentially as a result of differential reproductive success are responsible for the behavioral trait in question.

In the case of cultural learning, the step of selection for increased reproductive success may be severed from the step of transmission necessary for the evolution of the selected trait. No matter how greatly the trait enhances genetic reproductive success, that factor may have no role in the presence of the trait in future generations. Transmission follows a different path. Indeed, not only is increased fitness not sufficient for explaining the presence of the trait, it may not be necessary. (For a detailed argument, see Boyd and Richerson 1985; Brandon 1985b; Mitchell 1987a.)

When we observe or presume changes in behavioral traits, the inclination to treat those changes as adaptively significant requires strong assumptions about both reproductive consequences and the pathway from genetic replicators to behavior. In sum, for an individual behavior to be an adaptation it must be the direct cause of differential reproductive success and it must be directly transmitted as a result. In the case of complex strategies (the one-many relationship), an individual component behavior is not the unit of adaptation. Since adaptations involve both selection and evolution, the correct unit is the complete strategy, for it is the unit that gets transmitted as an integral whole. In the case of a learned behavior, it is the proximate learning mechanisms themselves that have been the object of evolution, and not any one or complex set of behaviors. Having identified the criteria for a unit of adaptation, one can then garner direct evidence that a given behavior (or behavioral strategy) is in fact an adaptation. Artificial selection experiments may be designed to justify the genetic underpinnings of a behavior, and field and lab studies used for determining the effects on reproductive success. Appropriate alternatives (either individual behaviors, strategies, or proximate cultural learning mechanisms) may then be proposed and discerned.

Classifications of Similarity

Up to now I have been concerned with what counts as an individual behavioral trait. Once the unit of adaptive behavior has been clarified, a derivative problem arises in comparing behaviors in one population or species with those found in another. This issue becomes especially relevant in the employment of the comparative method. The new question is, What are the criteria for similarity across populations or species or even taxa that justify ascribing the same name to two behaviors? That is, what are the criteria for similarity that allow the same evolutionary explanation to be inferred? The problems surrounding the link between behavior and genetic determinant might be hidden if the descriptions of significant social behaviors are cavalierly attached to correlative genetic replicators. For behaviors to be explained by their adaptive significance, the very same item that results in relatively higher reproductive success must be present because of that consequence. Sociobiological explanations often leave this presupposition ungrounded.

Burian argues that one of the problems with establishing the genetic determinism of social behaviors arises, in part, from the use of two different descriptive paradigms or "units of behavior" (Burian 1981: 53). We describe social behaviors from the perspective of their human significance using terms such as "aggressiveness" or "rape." But these do not necessarily correspond directly to the genetic units of behavior, namely, particular genetic replicators or gene complexes that both operate as coherent wholes in transmission and are responsible for the development of the behavior. Since adaptation explanations require the operation of both selection of phenotypes and transmission of corresponding replicators, there needs to be an account of how socially significant behaviors are mapped onto biologically significant units.

The blurring of descriptive categories is most likely to occur in indirect sociobiological arguments. The "comparative method" is often employed in order to justify sociobiological explanations of human behavior. That is, the "same" or similar behavior is studied in a variety of species that share certain environmental or structural similarities. Evidence is obtained for explaining the behavior's presence in one (or some) species and, by analogy, is extended to account for the presence of the "same" behavior in humans (Kitcher 1985: 184).

The Case of Rape

An example of the illegitimate grouping together of disparate behaviors under one descriptive category is found in Thornhill's work on rape in scorpion flies and humans (Thornhill 1980, 1984; Thornhill and Thornhill 1983). Thornhill's initial study of "rape" in scorpion flies suggested a generalized hypothesis about rape as an adaptive copulation strategy in any population where females exhibit choice of mates and males can secure material assets. He observed "rape" in the scorpion flies and tested to determine the reproductive consequences of the behavior as well as describing the associated ecological conditions. On the basis of direct investigation of the different copulation behaviors of scorpion flies (including copulation with and without the presentation by the male of a "nuptial gift" of a dead insect or salivary mass to the female), combined with information gleaned from studies of territorial fish and mallard ducks, Thornhill formulated a general hypothesis about the conditions under which "rape" as a copulation strategy would evolve by natural selection in any population. Humans were explicitly included in the scope of the hypothesis. He later (Thornhill and Thornhill 1983) was involved in a more direct consideration of the evolutionary significance of human rape; that is, he tested the predictions of the hypothesis he developed from his scorpion fly studies. To endorse a general hypothesis and be compelled to further test applications of it for humans presumes that "rape" in scorpion flies and "rape" in humans is a similar behavioral strategy, that the necessary ecological conditions are shared, and that the causal history that generated "rape" in each case would be the same and hence has the same adaptive significance for both.

Is it uncontroversial in the case of the flies that their behavior is genetically controlled, and hence explicable by consequences on reproductive success? The "unit" of behavior question is not entirely straightforward even in this case. No individual male adopts a single copulation behavior obligately. Rather, the evidence from Thornhill's own experiments is that any male will "rape" under the correct triggering environmental conditions. Similarly, any will, under appropriate conditions, send out a long-distance pheromone when either guarding a dead insect or after producing a salivary mass. Hence, "rape" is not an isolated behavioral alternative subject to selection, but rather part of a complex strategy for copulation that includes a set of outcome behaviors that depend on environmental triggers. In short, it is a component of a conditional male reproductive strategy composed of behavioral alternatives. It might be better represented as the ordered sequence:

1. If possessing dead arthropod, emit long-distance pheromone, then copulate.
2. If no dead arthropod, produce salivary mass, emit long-distance pheromone, then copulate.
3. If no dead arthropod and unable to produce salivary mass, and female is present, secure female with physical force, then copulate.

Understanding it thus, it is clear that identifying "rape" as an adaptation is shorthand for claiming that the complex conditional strategy that includes "rape" is adaptive. One must then make a case for variant copulation strategies being present in the evolutionary history of the scorpion flies such that some included the "rape" component and others did not. It should be pointed out that in more recent writings (Thornhill and Thornhill 1983, 1987) the conditional, facultative nature of copulation strategies is explicitly acknowledged. However, the evolutionary hypothesis is still framed in terms of *rape* being adaptive or maladaptive.

Clarifying what are alternative strategies subject to evolutionary explanation is important in the context of game-theoretic analysis as well. Those studies focus on explaining the maintenance of behavioral variation in a population by means of frequency-dependent or disruptive selection processes. Austad has pointed out that confusion resulted from the absence of common terminology for describing behavioral alternatives. The use of both "tactic" and "strategy" indiscriminately in referring to behavioral components of complex strategies and obligate behaviors as well as the complex strategy as a whole produced the ambiguity (Austad 1984). Clearly, if the unit of behavior is ambiguous, identifying alternative behaviors will be hopeless.

Suppose the evidence is convincing that copulation strategies that include "rape" are adaptive for the scorpion flies. Can we then infer that human rape is similarly explained? For an analogical argument to support an explanation of human behavior as adaptive, the properties appealed to as shared must be relevantly similar. Serious doubts can be raised as to the successful grouping together of human behavior and fly behavior as "rape." If this similarity fails, then the evidence that the behavior in the one species is adaptive lends little credence to the claim that it is so for the other species as well.

Let us look more closely, then, at the identification and explanation of "rape" in scorpion flies in order to see whether evolutionary explanations of its adaptive nature are justified there, and if so, whether they can be extended to explain rape in humans as well. Thornhill describes the observable sequence of events that constitute "rape" in scorpion flies:

> A rape attempt involves a male without a nuptial offering (i.e. dead insect or salivary mass) rushing toward a passing female and lashing out his mobile abdomen at her. On the end of the abdomen is a large, muscular genital bulb with a terminal pair of genital claspers. If the male successfully grasps a leg or wing of the female with his genital claspers, he slowly attempts to reposition the female. He then secures the anterior edge of the female's right forewing in the notal organ.... Females flee from males without nuptial gifts. If grasped by such a male's genital claspers, females fight vigorously to escape. When the female's wings are secured, the male attempts to grasp the genitalia of the female with his genital claspers. The female attempts to keep her abdominal tip away from the male's probing claspers. The male retains hold of the female's wing with the notal organ during copulation, which may last a few hours in some species. (Thornhill 1980: 53)

Is what is described an instance of "rape"? To demonstrate that it is, Thornhill suggests two criteria that must be met: "... it is necessary to (1) clearly distinguish female coyness from rape and (2) show that males that rape enhance their own fitness" (Thornhill 1980: 52). What is the motivation for these criteria? The descriptive content of the term "rape" must be derived from its use in human social contexts. For humans, purely behavioral information is insufficient to determine that a social interaction counts as rape (throughout, "rape" means only heterosexual rape). There is an essential intentional component. For it to be rape, two psychological conditions must be true: The female must be unwilling to engage in sexual intercourse, and the male must be willing. No behavioral expression is either necessary or sufficient to characterize this behavior.

This definition is unassailable if you consider a set of behavioral observations and ask yourself whether any count as cases of rape. Consider a case of copulation where there is physical struggle between male and female. On the face of it, this can be either consenting sadistic-masochistic behavior or it could truly be rape. What makes the difference is the intentional attitudes of the participants, not the behavioral counterparts. What if there is no physical struggle associated with intercourse? That behavioral set does not guarantee that what is going on is not rape. Fear of physical harm induced by threats can easily account for cases when the female is not consenting and yet not physically struggling. Again, what makes it genuine rape has to do with intentional attitudes. Since all such cases are male-induced actions, the assumption of male willingness is unproblematic.

It is crucial to separate ontological from epistemological or evaluative judgments. The legal definition of rape involves assigning culpability and degree of punishment appropriate to a given case. Here the kind of behavioral evidence I have claimed to be inessential in defining rape may well come into play. But this is not a question of whether or not the action *is* rape, but whether or not the parties involved had good evidence for knowing that. For purposes of evaluation, one must first be able to identify the behavior.

How do the criteria for human rape correspond to the criteria used in Thornhill's study? To apply a concept such as "rape," which is essentially intentional, not just correlated with intentions, an analog to unwillingness must be found. Thornhill's first criterion, distinction from female coyness, presumably plays that role. But what is packed into this distinction? Female coyness is taken to be a way for the female to exercise a discriminating role in interactions. By not engaging immediately in copulation behavior, the female may be able to elicit information about the male's fitness. Thus, in the case of the scorpion fly behavior Thornhill described above, it is plausible that the female, by struggling to free herself from the grasp of the male, is in fact determining whether he is strong enough to hold her, and hence likely to have those features that make him well adapted to a hazardous environment. Thornhill points out that 65 percent of adult mortality is due to predation by web-building spiders (Thornhill 1980: 54.). If the female is using the struggle to evaluate male fitness, then the behavior described is not "rape." What Thornhill argues, however, is that given his studies, which show an ordered preference of females for males with large, rather than small, dead arthropods as nuptial gifts, and dead arthropods rather than salivary masses, there is no reproductive advantage for the female to copulate with a male who fails to present a nuptial gift (Thornhill 1984: 91). To be fair, Thornhill has refined his hypotheses regarding scorpion fly preferences, testing the contributions of body size, prey size, and frequency of males on female choice (see Thornhill 1984 for a list of his relevant studies). By receiving a gift, the female acquires nourishment that otherwise she would have to obtain by means of risky foraging in a hostile environment. Both the fact that a male offering a gift or salivary mass displays his ability to acquire food in that environment (producing a salivary mass is possible only after a male has recently fed) and the fact that she directly benefits materially from the food make copulation with a "rapist" less beneficial. So to defend the view that the behavior is "rape," Thornhill must presuppose that to distinguish it from coyness the behavior must be clearly *not* in the reproductive interests of the female. Thus the analog to unwillingness is reproductive disadvantage.

This interpretation is consistent with the other half of the set of criteria. A male's fitness is enhanced, so Thornhill argues, by inseminating a female without having to put himself at risk foraging to acquire a dead arthropod to present or as a means of producing a salivary mass. What makes the incidence of "rape" in the flies so rare is the lower frequency of successful insemination for that behavior. Thus, for a behavior to be "rape," it has to be, at the same time, in the reproductive interests of the male and against the reproductive interests of the female.

There are two criticisms to be raised to this analysis of "rape" in scorpion flies. The first is that it is not clear that the two criteria that Thornhill has set out are met. That is, it is not clear that the behavior is really "rape." The behavior of the female in struggling to free herself from the grasp of the male has not been sufficiently distinguished from coyness. Her behavior may elicit just the sort of fitness information she requires to make copulation in her reproductive interests. Furthermore, given the aggressive nature of the interactions with conspecifics, heterospecifics, and predators in that environment, the female's behavior may have nothing whatsoever to do with her willingness or unwillingness to copulate with a given male. Struggle, rather than indicating an instinct to avoid copulation, may be part of a different behavioral set. The female may be avoiding being grasped in general, a behavior much to her advantage. Supporting this view are the following facts: There is no pheromone release from the "rapist" male; pheromones are used for species recognition; and there is aggressive behavior over food between conspecific males, between heterospecific pairs of any sex, and with predatory spiders (Thornhill 1984: 81–83). Given this, it is plausible that the female's struggle with the "rapist" has nothing to do with copulation at all.

The second criticism is more global. If, in fact, the criteria are met, and the behavior is thus classified as "rape," by so doing Thornhill cannot ask the further question, Is "rape" adaptively significant? That is because to be rape in the first place, by definition it entails that the behavior is in the reproductive interests of the male and contrary to the reproductive interests of the female. A more neutral classification of the behavior would allow the question of adaptive significance to be raised, tested, and disputed.

With respect to the extension of the adaptive significance of "rape" in scorpion flies to rape in humans, two points must be made. The first has to do with the analog of unwillingness. In the case of humans, unwillingness refers to some proximate psychological mechanism that, while itself having an evolutionary history, can generate behaviors that are not directly subject to the process of evolution by natural selection. That means that, for humans, unwillingness is not directly correlated with consequences of reproductive failure. This is just the point of the distinction drawn above between behaviors that are part of a closed behavioral program and behaviors that are generated by cultural learning mechanisms. The relevant causal processes for these two categories are distinguishable. For that reason, the adaptive significance for what we may describe as "rape" in one case cannot be generalized to cover a behavior with a different, and not necessarily complimentary, causal history in the other case.

This slide between components of closed and open behavioral programs is made by those who believe that because a cultural learning program has itself evolved by means of natural selection, particular behaviors proximately generated by it must necessarily be in the reproductive interests of the actors. Thornhill and Thornhill (1983: 139), while recognizing the different causal routes involved in the evolution of human behavior, that is, evolution via natural selection and learning via cultural selection models, nevertheless insist that "[b]oth . . . routes are expected to result in behavior that promotes inclusive fitness [number of descendent and nondescendent kin] of individuals" (see also Durham 1979; Irons 1979; Shields and Shields 1983). However, Boyd and Richerson have convincingly argued that there is a consistent evolutionary account of how a cultural learning model can arise by natural selection and yet generate behaviors that do not promote the genetic fitness interests of the actors who express them (Boyd and Richerson 1985; Brandon 1985b; Mitchell 1987a). Once that inexorable relationship between behaviors generated by a proximate learning mechanism and consequences that increase the reproductive fitness of actors is given up, then all the statistical evidence that Thornhill and Thornhill provide on the reproductive consequences for rapists and raped women are beside the point.

Sociobiologists attempt to explain human behavior as biologically adaptive. Why a social behavior is present in a specific human society is answered by appeal to the consequences of that behavior for maximizing inclusive fitness. To be adequate to the task, such explanations require that there be an appropriate connection between the causal background processes. While it may be shown that a behavior confers a relative advantage for reproductive success onto the actor, for the behavior to be an adaptation, that advantage must be the cause of the behavior being present in the population. Such is the case when a trait is genetically determined, since the advantage in reproductive success directly corresponds to an increase in genetic replication. However, when the direct connection between behavior and genetic replicators is severed, then the causal significance of the consequence of a behavior on reproductive success becomes questionable.

Whereas some types of behavioral variability are indeed the result of evolution by natural selection, namely, those that are generated and transmitted by genes, other types of behavioral variability have a different etiology. The consequences on inclusive fitness of behaviors of the first type of variability can indeed explain why such behaviors are present in a population, even though there is no direct link between genetic replicator and specific behavior. The consequences on inclusive fitness will not, however, explain why a behavior is present that is learned and transmitted by imitation. Neither direct nor indirect arguments for the adaptive significance will help explain behaviors that are the result of a causal history independent of the combined processes of natural selection and evolution. (Thornhill and Thornhill's most recent collaboration (1989) shifts the focus of explanation of human rape to the psychological mechanisms generating male rape behavior and away from the behavior itself.)

In launching an investigation into the adaptive significance of a trait, it is advantageous to clearly identify the target of evolution by natural selection. Making explicit the assumptions concerning the relationship of the target to the genetic substrate, the specification of phenotypic alternatives, and the environmental conditions and reproductive consequences of the alternatives will aid in avoiding the comparison of incommensurable traits across species. Inferences of the same adaptive significance for similar targets of selection involve justification that the same causal forces are at work. Ultimately, it is the causal history of the trait that determines whether it is an adaptation and the nature of its adaptiveness.[12]

4.3. ON BIOLOGICAL FUNCTIONS

There has been a recent resurgence of interest in functional explanations, both in the details of the analysis of functions and in the appropriate strategy for carrying out such an analysis (Bigelow and Pargetter 1987; Horan 1989; Millikan 1984, 1989; Mitchell 1989; Neander 1991).[13] Some, including myself, have defended and developed an etiological theory of proper functions, whose origin derives in part from the work of Larry Wright (1973, 1976). On this theory, a function is a consequence of the presence of some component of a system: The pumping of blood consequence of the human heart, the inclusive fitness maximizing consequence of the avunculate in societies with low paternity certainty, the keeping open of the door consequence of the brick on the floor. The consequence, to be a function, must have played an essential role in the causal history issuing in the presence of that very component. While agreeing on the basic character of functions, the defenders of etiological theories nevertheless disagree with respect to the nature of their enterprise, that is, whether it is an act of conceptual analysis or one of theoretical definition. Clearly, the type of arguments available to claim success for the etiological theory, and its right to prevail over other contenders in the field, depend on what such accounts aim to do. Prior to entering some new battlegrounds in the war between function theories, I first address this meta-analytic question. I argue that current disputes may be better understood, if not dissolved, by understanding the explanatory enterprise in which appeal is made to functional ascriptions. In this pursuit I consider two

---

[12] I thank the following people for helpful discussions of the topics in this section: James Bogen, Michael Dietrich, Jerry Downhower, Philip Kitcher, Peter Machamer, and Rob Page. Thanks also to Randy Thornhill for sending me his recent manuscripts.

[13] This section was originally published as two separate papers: "Dispositions or Etiologies: A Comment on Bigelow and Pargetter," *Journal of Philosophy* 90, no. 5 (May 1993): 249–259, and "Function, Fitness and Disposition," *Biology and Philosophy* 10 (1995): 39–54. Redundancies were removed.

recent criticisms of the etiological theory brought forward by Bigelow and Pargetter, and the relative merits of their alternative, dispositional theory of functions.[14]

Setting the Stage

Functional claims are endemic to biology and the social sciences. "The function of the heart is to pump blood." "The function of melanic coloration is camouflage from predators." "The function of the Hindu taboo on eating cattle is to provide traction animals for subsistence farming." While occurring in a variety of contexts, function claims are clearly intended to do explanatory work. However, the teleological character of functions, namely, the fact that a function is identified with a consequence of the presence of the very item it is supposed to explain, has been taken as prima facie grounds for denying them explanatory status. This methodological criticism was taken as evidence of the gulf between the natural sciences and the social sciences.

Hempel (1965b) attempted to restore the unity of the sciences under the methodological rubric of the covering-law model of explanation. On this account, which had become the "received view," a functional explanation of why a given trait or practice is present in a system consists in its derivation from a set of premises that include a statement of the function of the trait, and laws and initial conditions. In the premises are statements both that the system is operating adequately and that a necessary condition for it to do so is fulfillment of the function. The laws of biology and social science, however, do not permit the derivation of claims such as a heart is present, or melanism is characteristic in a population of moths, or the Hindu food taboo is practiced, because they cannot rule out functional equivalents for these explanandum events. Mechanical hearts, behavioral concealment, and conscious social planning would do equally well in achieving the functional consequences issuing from human hearts, camouflage coloration, and food taboos. On this analysis, while functions may indicate interesting relations between the parts of a system and its overall operation, they fail to explain why a particular part is present. For this reason, Hempel allocated to functions only a heuristic role in scientific discourse.

Hempel's negative solution to the philosophical puzzle engendered a new methodological problem. Biologists and anthropologists appeal to functions

---

[14] Bigelow and Pargetter 1987.

in ways that assume more than mere heuristic content. They dispute what is the correct function to ascribe to an organ, behavior, or social practice. Empirical evidence is brought to bear in identifying functions. In evolutionary theories, functions are associated with adaptations and seem to be explanatory in just the way the Hempelian analysis precludes. For example, G. C. Williams states: "Evolutionary adaptation is a special and onerous concept that should not be used unnecessarily, and an effect should not be called a function unless it is clearly produced by design and not by chance."[15]

Hempel's solution rules that scientific disputes about the correct attribution of function are misdirected, and that scientific language and practice must be reformed in light of the philosophical analysis of function. However, some have responded to the mismatch between the Hempelian analysis and the explanatory uses of function by taking up the challenge to revise the philosophical analysis in order to account for the explanatory character presupposed in scientific practice.[16]

But what type of philosophical enterprise is this? What do we want from a theory of explanatory functions, and hence what constitutes evidence that we have achieved our aims? Millikan (1989) defends an etiological theory of proper functions as a theoretical definition and rejects Wright's similar theory (1973, 1976) on the grounds that his is just conceptual analysis. Neander (1991), in response to Millikan's argument, distinguishes weak and strong interpretations of conceptual analysis. In the weak version, the conceptual analysis of the term "function" would identify implicit or explicit criteria of application for its utterance in the appropriate linguistic community. A strong interpretation adds to this the requirement that the analysis provide an account of the meaning of the term, and that it do so by outlining necessary and sufficient conditions for its use. These additional requirements are what Millikan rejects and motivate her promoting a theoretical definition of "function" and at the same time eschewing conceptual analysis. A theoretical definition aims not at an account of the meaning of a term, but rather at describing the underlying phenomena.

Neander is correct to defend the weak version of conceptual analysis here and to suggest that it be used in conjunction with a theoretical definition. However, I believe there is a stronger reason for adopting a combined approach than Neander's appeal to the "related aspects of language" that each approach illuminates. That is, "function" does not pick out a material substance like the natural kind terms "gold" or "water," which serve as exemplars for comparing

---

[15] Williams 1966: vii.

[16] See Bigelow and Pargetter 1987 for an account of the range of such views in the current market.

conceptual analyses with theoretical descriptions for Millikan and Neander. It is more like "cause" or "reason" because it is both abstract and picks out an explanatory structure.[17]

Understanding the intended explanatory content of the use of "function" in current science will indeed employ tools of weak conceptual analysis (since it is the scientists' term and not a stipulation outside science that is at issue) and at the same time provide a theoretical definition that grounds its "correct" use in contemporary science. Rather than ask what scientists mean by "function" or what objects the term denotes in the world, I suggest we explore instead the explanatory import of function ascriptions. That is, we should ask *what* they are intended to explain and *how* they explain it. This approach will clarify both the content of the etiological theory and its differences with the dispositional alternative offered by Bigelow and Pargetter.

The etiological approach maintains that to explain why something occurs is to describe the causal history that led to the event, to give its etiology. The plausibility of any posited function derives from the theories in our scientific repertoire. But acceptance of a given ascription is subject to the same evaluation as any other scientific claim.[18] In short, which history is appropriate is context-dependent. Which history is true depends on the world. While all theories of scientific explanation might well agree that causes are explanatory, functions are challenging since they are not causal in the usual sense, they are teleological.

There have been many accounts of how intentional, goal-directed behavior is appropriately teleological (Cummins 1975; Nagel 1977; Wimsatt 1972; Wright 1976). The idea or the representation of the goal of an action in the mind of the actor operates as the proximate cause for that same action. My concerns, however, are with the explanation of systems to which the assignment of goals or intentions may not be appropriate. Can the presence and persistence of properties of biological populations and human societies be explained by appeal to natural or cultural functions? Larry Wright and Ruth Millikan have offered such accounts.

Wright (1976) proposes the following schema:

The function of *X* is *Z* if:
i) *Z* is a consequence (result) of *X*'s being there
ii) *X* is there because it does (results in) *Z*.

---

[17] Or a "methodological" concept, as Nordmann suggests, which needs to be "worked through again and again" (1990: 380).

[18] Or, if you prefer, which function ascription is accepted by the scientific community will depend on the constitutive standards for evidence and confirmation in that community.

Satisfying clause (ii) of the formula, that X is there because it does Z, depends on the appropriate causal relationship obtaining between X and Z. A rough version of Millikan's (1989: 288) theory is:

> A has F as its proper function if one of two conditions hold:
> i) A originated as a "reproduction" of some prior item or items that due in part to possession of the properties reproduced, have actually performed F in the past, and
>
> A exists (causally, historically because) of this performance or
>
> ii) A has F as a derived proper function, i.e. A originated as the product of some proper device that had performance of F as a proper function and that production of A is the normal means by which F is achieved.

Neither Wright nor Millikan makes explicit the specific characteristics of the historical, causal "because" that carries the explanatory burden in their accounts. Clearly, not all causal sequences render a consequence a function. To do so two conditions are required: first, that the organ or behavior has been selected over alternatives on the basis of its consequence, and second, that it is produced or reproduced as a direct result of that selection process (Mitchell 1989, 1993). Wright hints in this direction when he says that "functional explanation depends essentially on a selection background" (Wright 1976: 101).

Natural Functions and Social Functions

Given this characterization of the two components of an etiological theory, it is clear how the etiological theory applies to natural items when their presence is the result of evolution by natural selection. The presence of traits that are adaptations can be explained by appeal to their past consequences on reproductive success. For example, in primates, males are often larger than females. What function does larger size serve for the males? Two answers were proposed. Darwin (1965) said that larger size in males was a result of sexual selection, because larger size resulted in greater success in male-to-male competition for mates. Selander (1972) suggested that the dimorphism allowed males and females to exploit different food resources. Clutton-Brock and Harvey (1977) tested the two hypotheses with comparative data. If the function of larger size for males was for food acquisition, then one would expect to find this dimorphism in monogamous species of primates where males and females feed together. Here the dimorphism would be due to the consequence of larger size in competition with females for food. If the function were, instead, to enhance success in male-to-male competition for mates, one would expect to find dimorphism in polygamous species. In these situations, there is strong competition between males for access to mates where the winner acquires a harem of females and the loser none. Clutton-Brock and Harvey's study gave evidence supporting Darwin's hypothesis for primates. Dimorphism appeared in polygamous species. What is evident from this discussion is that the biological function of a trait is taken to be a real, discoverable property of natural organisms. Determining the correct function requires empirical evidence about conditions under which selection could occur and hence the consequences that the trait had in the past.

For an etiological functional relationship to occur, a selection background such as natural selection must operate. The selected consequence of an item must furthermore be causally responsible for its replication and, hence, its current presence. Evidence only of a selection background or selection *of* an item will be insufficient to justify ascription of a function. The feature explained, such as larger size in male primates, must have been selected *for* the functional consequence, say, success in male-to-male competition, and, second, it must have been produced or reproduced as a direct result of that selection process, in this case by genetic transmission of a heritable trait.

Function is also used to explain the presence of social practices. This can be understood by an etiological theory only if a selection process is operating that takes the alleged functional consequence as causally relevant in the maintenance or transmission of cultural items. Functionalism in anthropology and sociology has been often associated with the program to explain human behavior and social institutions in terms of group-level functions. That is, specific beliefs, behaviors, and institutions exist because they allow the healthy functioning of the social group. Rituals maintain social cohesion, warfare regulates population size, and marriage rules maintain cross-cultural social ties. Social practice X is there because it does Y, that is, contributes to the health or welfare of the social group.

A classic contribution to ecological anthropology has been the research by Rappaport and Vayda on the Maring warfare cycle (Rappaport 1968; Vayda 1974). The Marings of Papua New Guinea ritually plant yams and hold pig feasts in a cycle of war and truce making with neighboring populations. It has been claimed that the function of this complex set of behaviors is to distribute protein to allies during the pig feasts and regulate population size by means of the victors confiscating the land of the defeated group. These consequences occur within the context of the existing modes of production and environmental constraints.

A common criticism of functional explanation in cultural anthropology is taking the fact that a consequence is beneficial as sufficient to accounting for why the practice exists. Indeed, this is called the "fallacy of function-alism." The etiological theory of function makes sense of such criticisms. Without evidence of selection and transmission mechanisms operating on the consequences of a behavior to further generate the behavior, ascription of etiological function is groundless.

There has been a variety of responses to the recognition of this fallacy. Some, including Vayda, have taken seriously the requirement for causal mechanisms for etiological functions and have redirected their research toward gathering such evidence.

> Contrary to the assumptions made by me in the 1970s and by other anthropologists more recently, the fact that victorious warriors sometimes take enemy land and benefit from doing so has, by itself, no necessary explanatory import. To think that it does is a fallacy, not because it involves putting consequences forth as causes, but rather because it involves putting them forth without due concern for mechanisms, as if the mere fact of their being beneficial automatically conferred causal efficacy upon them. The mechanisms to which I am referring are . . . intentional action, reinforcement, and natural selection, ontologically grounded in the actions, properties, and experiences of individual human beings. (Vayda 1989: 163)

Just acknowledging the need for selection and replication mechanisms is not in itself sufficient. Vayda cautions against a "ceremonial" invocation of mechanism without any investigation into its operation. Additional evidence is required for identifying the actual selection and replication processes from the set of those that could logically do the job. Vayda now claims that conscious recognition of the consequences and intentional action is responsible for the Maring ritual warfare cycle. To support the claim that redistribution of land is indeed the function of specific war cycles in the 1950s, Vayda has documented the fact that there was land shortage at the time, and that individuals consciously acted on the basis of this fact.

Individual intentional action is not the only selection and transmission mechanism available to support cultural functions. Soltis, Boyd, and Richerson (1991) raise the question of whether functional accounts of cultural behavior can be justified by means of cultural group selection. They develop a model that suggests that under certain circumstances, cultural variation can be maintained and "cultural group selection could cause human societies to exhibit at least some group functional traits" (1991: 4). Group dissolution or extinction, and fissioning combined with cultural transmission, are mechanisms that might explain the presence of practices such as the warfare rituals in Papua New Guinea.[19] Groups that have a given practice and that are successful at survival or replication because of it may pass this practice on through their lineage and to their offshoots by cultural transmission. Soltis, Boyd, and Richerson present evidence from New Guinea that supports the realism of the assumptions of their model, though it is not sufficient for fully justifying the claim that group selection actually operated in that case.

Just as in the case of natural function, an etiological explanation for cultural function requires detailed, context-specific evidence of the actual causal history. Not every beneficial consequence will be explanatory, and not every explanatory consequence will invoke one particular mechanism, be it intentional choice, natural selection, or cultural group selection. What the etiological theory accounts for is the structure of cultural functional explanation, whichever selection and replication mechanisms are in evidence.

To summarize, what is crucial to functional explanation is that a specified type of causal background is operating and that the functional consequence played the appropriate role in the causal history leading to the trait. Two types of evidence are required: having the trait issued in such a consequence in the past which was responsible for the selection of the trait over others, and the selection drove the process of its production or replication. The etiological analysis thus can show why functions are explanatory and how they increase our scientific understanding of the presence of traits by indicating the causal history responsible for their presence. Some practices of biologists and social scientists, including theoretical development, empirical investigation, and dispute resolution, make sense only if an etiological theory of function is presumed.

Malfunction and the Problem of Doubles

Two intuition-driven counterexamples have been raised against the etiological approach. The first, the problem of accommodating malfunctioning items, such as congenitally diseased hearts, has been heralded, curiously, as both a problem for an etiological account by Prior (1985) and an advantage of an etiological account by Millikan (1989). The problem that Prior identifies rests in how stringently one interprets Wright's condition (i) that Z is a consequence (result) of X's being there. If one insists that each individual X must result in Z if Z is a function of X, then malfunctioning hearts do not really have the function to pump blood, since not only do they fail to, but they cannot succeed under any circumstances. This distinguishes them from things such

---

[19] See also Boyd and Richerson 1991.

as safety devices that may not at any given time result in Z but would if the proper circumstances obtained. The airbag in a car may never be used, but its function is nevertheless to protect the driver from the impact of the steering wheel should a collision occur.

If one adopts an interpretation of (i) to allow Z to be a typical consequence or result of the presence of things of type X, then the problem with malfunction dissolves. If things identified as Xs typically result in Z and Wright's condition (ii) is met, then an individual X that fails to result in Z, the typical consequence of things of types X being present, is deemed malfunctional. Millikan defines biological categories in terms of functions. They are things that are grouped together in virtue of their causal history. An individual member of a category, of course, may be defective and hence unable to produce the consequences that define the category, but that does not diminish its membership. A heart that fails to pump blood is still a heart.

> It is not then the actual constitution, powers, or dispositions of a thing that make it a member of a certain biological category. My claim will be that it is the "proper function" of a thing that puts it in a biological category, and this has to do not with its powers but with its history. (Millikan 1984: 17).

The objection that malfunction cannot be identified in an etiological theory is thus met by distinguishing types from tokens and by adopting a nonessentialist interpretation of types. Especially for biology, where traits and groups are both variable and evolving, there will be no set of necessary and sufficient properties that must be expressed by every item classified as a heart or a Homo sapiens. What allows the identification of a token as being of a specific type is historical continuity. What is the "normal" or "proper" function of that type is understood in terms of potential consequences the expression of which is environmentally conditional.[20] Thus, malfunction can be ascribed to items that meet the historical identity conditions for type classification but fail to have "normal" expression in a given environment.

The second counterexample to etiological theories of function is the problem of doubles. Bigelow and Pargetter suggest the following thought experiment to illustrate:[21] "Consider the possible world identical to this one in all matters of laws and particular matters of fact, except that it came into existence by chance (or without cause) five minutes ago" (1987: 188). Allegedly, the etiological account would find none of the items we identify as functions in our world to be functions in the parallel world. Clearly, in the parallel world,

---

[20] See Hull 1987; Millikan 1989: 300.
[21] See also similar cases given by Prior 1985.

by design of the example, there is no causal history in which to embed them. The problem of doubles offers two objects with identical current properties but with different histories. Bigelow and Pargetter claim our intuitions dictate that if the items are identical in structure and in all future consequences and one has function X, the other must have the same function.

Millikan, in response, objected that they mistake a mark of a function, its structure and its current consequences, for the function itself, the causal historical significance of the structure/consequence pair.[22] While I agree, I find Millikan's argument to not be very satisfying. She admits to being brazen in just stating that the cases of doubles are like cases of gold and fool's gold, or H_2O and XYZ, cases that look alike to us, but are simply mistaken identities. This rather begs the question, since for those who take the doubles case as persuasive, any adequate account of function should construe doubles as functionally identical. For Bigelow and Pargetter it is Millikan who has made the mistake by seeing doubles as functionally different. However, intuitions about fictional worlds, on either side, seem poor grounds for determining allegiance to one or another theory of functional explanation. If intuitions are to play a role, I believe they should be intuitions about our world.

At least for some evolutionary biologists, to ascribe a function to a trait does seem to require a causal history of adaptation. Recall Williams's declaration that ". . . no effect should be called a function unless it is clearly produced by design and not by chance" (Williams 1966: vii). This perspective, if applied to Bigelow and Pargetter's double world, would see traits but no functions, since in that case chance produced the traits.

I suggest it might be helpful to consider what intuitions are evoked by considering a natural double. In the case of mimicry, members of two distinct species display very similar structures or traits, in fact indistinguishable to the appropriate parties, as well as similar current consequences of these traits, and yet the populations have experienced different causal histories. Do we say that the traits of models and mimics, the natural doubles, have the same function in line with what Bigelow and Pargetter do for the unnatural doubles?

Take the case of the monarch butterfly (*Danaus plexippus*). These butterflies, found in North America, are brightly colored, orange and white or black and white, and are abundant. An early puzzle for Darwinism was explaining such conspicuous coloration. After all, one could see that camouflage coloration would be adaptive by helping an organism avoid predators.

---

[22] Similar objections to adaptation claims are made by Gould, who criticizes biologists who mistake current consequences for traits as sufficient evidence for the ascription of adaptive function. For discussion of this argument, see Alcock 1987; Gould 1987a, 1987b; and Mitchell 1992.

But conspicuous coloration presents a detectable visual signal to a predator. It turns out that having easily identifiable markings may have evolved for the function of signaling unpalatability. The monarch caterpillars feed on milkweeds and thereby absorb and store cardenolides, a distasteful substance that makes them inedible both as caterpillars and as butterflies. If predators learn that monarchs are unpalatable, then they will not eat them. Experiments in the laboratory and in the wild indicate that it is indeed the case that birds have aversive reactions to monarchs after having tasted one, and they decrease the frequency of biting them until completely stopping, thereby substantially reducing predation rates.[23] The body of experiments has been taken as evidence that "the function of conspicuous color patterns is aposematic [i.e. warning of unpalatability]".[24] Now consider the viceroy butterfly (*Nympha/lidae, Limenitus archippus*), which is the natural double to the monarch in color pattern, even though it belongs not only to another species, but to a different family. The viceroy has a diet unlike the monarch's, and is indeed a desirable food source for birds. Nevertheless, the conspicuous coloration in the viceroy also allows it to avoid predation, by being a mimic of the unpalatable monarch. In cases of Batesian mimicry in butterflies, the operator or predator cannot visually distinguish individuals from the two unrelated prey species. One organism, the model, is both conspicuously colored and unpalatable. The other, the mimic, is similarly colored, but palatable. The evolution of Batesian mimicry requires that the model organism must be more abundant than the mimic or the learning process would work against the association of conspicuous coloration with unpalatability. Studies of models and mimics have shown that mimetic resemblance tracks changes in morphology of the model, disappears in the absence of a model, and is controlled by a complex of genes.[25]

I have presented what I take to be a case of natural doubles. Two types of organism, the monarch butterfly and the viceroy butterfly, are structurally similar or indistinguishable. Parallel to Bigelow and Pargetter's double-world example, the morphological structures have the same future consequences, that is, avoiding predation, but have had different evolutionary histories. Do we want to say that the conspicuous colorations of the monarch and viceroy have the same function? No. Traits of mimics and models are not the same. As Wickler (1968: 108) says: "In general, we use the term mimicry only when the mimetic characters have been evolved for a specific mimetic function."

---

[23] See Brower 1960, 1988 for experimental results.

[24] Guilford 1988: 9. Guilford argues, however, that this is insufficient evidence for the conclusion, and suggests various evolutionary explanations for the association between conspicuous coloration and unpalatability.

[25] Wickler 1968.

The function of conspicuous coloration in the monarch is to warn the predator of its unpalatability. The function of the viceroy coloration is to mimic the model and deceive the predator into presuming it is unpalatable and thereby avoid predation. The same structure has two functions: One is to warn and the other is to deceive.

Thus, if two butterflies flit into our view and we see virtually identical structures and virtually identical current consequences on predation avoidance, Bigelow and Pargetter might have us conclude that the features of the model and the mimic have the same function. But surely that is a mistake. By knowing the causal history of conspicuous coloration, the models and mimics can be distinguished, and the functions of the traits understood. Perhaps a more compelling argument could be made if two identical structures had more divergent evolutionary histories, that is, one the product of direct selection of the trait for the resulting consequence, and the other being a case of genetic hitchhiking such that it was not selected directly for any feature save its location on a chromosome. While this is perfectly plausible on current biological theory, I know of no actual case that displays sufficient similarity to be a natural double of this kind. Nevertheless, with the mimicry example, I hope to have shown that knowledge of identical current structures is not sufficient to compel assent to identical function. History does matter to biological function, and it is that fact that the etiological account acknowledges.

Bigelow and Pargetter grant that an etiological theory of function can render function explanatory, but at the cost of looking *backward* to causal history. The doubles thought experiment was intended to suggest the need for a theory of function that is not just explanatory, but also *forward-looking*. To satisfy this intuition they propose a dispositional theory of function, based on an argument by analogy with biological fitness.

Biological Fitness Analogy

Bigelow and Pargetter provide a more direct argument against the etiological approach and in support of their alternative, dispositional account of functions. They aim to replace the *backward-looking* etiological approach with a more *forward-looking* dispositional account of functions. To this end, they offer an intriguing argument by analogy between the concepts of biological function and biological fitness.

They propose to characterize functions not in terms of causal histories but rather in terms of future effects, that is, as dispositions. They ground this view on an analogy with a strategy for restoring explanatory import to another important concept, biological fitness. In short, Bigelow and Pargetter argue that just as the concept of fitness lacked explanatory status when it was identified with actual reproductive success and that status was restored by construing fitness as a disposition or propensity to have a certain degree of reproductive success, so, too, with the concept of function. A function fails to be explanatory, they maintain, when it is characterized solely as an actual consequence; similarly, its explanatory status can be saved by being characterized as the disposition to have that consequence. "Fitness is forward-looking. Functions should be forward-looking in the same way and, hence, are explanatory in the same way" (Bigelow and Pargetter 1987: 191).

There is something quite seductive about this argument. After all, the dispositional or propensity theory of fitness did solve an explanatory failure problem, and so might well be a powerful strategy to apply elsewhere. I argue, however, that on deeper reflection both the negative and positive claims of the analogy are faulty. Not only is there no corresponding tautology produced by construing functions in terms of actual consequences, but the move to a dispositional analysis shifts the very target of explanation. Thus, the dispositional theory of functions fails to address the problematic of the etiological view, namely, why the functional trait is present. I consider the negative and positive arguments in turn.

Bigelow and Pargetter claim that if functions are construed as actual past consequences of the presence of the functional trait, as the etiological theory would have it, then "[i]t is no longer possible to explain why a character has persisted by saying that the character has persisted because it serves a given function" (190). There are two ways to unpack this argument, only one of which results in a tautology. The proposed explanation (E) is:

(E) Trait X persisted because it served function Y.

Let Y be interpreted etiologically. Y is that consequence that was responsible for the selection of X over alternatives and the subsequent replication of X through successive generations. The etiological interpretation translates (E) into

(ET) Trait X persisted because it had a consequence responsible for its selection and consequent evolution.

A tautology arises only when one begs the question of what causal process was responsible for the persistence of X. That is, (ET) is a tautology (T) only when "persisted" is read as "evolved by natural selection."

(T) Trait X evolved by natural selection because it had a consequence that caused it to evolve by natural selection.

The original question, however, is why trait X persisted. One answer to that question is that it did so by a process of evolution by natural selection for a particular consequence, that is, the evolutionary functional answer. But there are other causes for the persistence of a trait, namely, chance or drift, physical necessity, or developmental constraint. Appeal to actual functional consequences in the past does make (ET) explanatory. The etiological function explains the persistence of X by specifying that one process, rather than other potentially active processes, was causally responsible. When function refers to actual, historically active consequences, explanation of persistence by appeal to function is not tautological. So much for the tautology claim.

The Disposition Argument

Bigelow and Pargetter's positive analogy consists of the claim that biological function and biological fitness are explanatory in the same way, and that is the way of dispositions. "Fitness is forward-looking. Functions should be forward-looking in the same way and, hence, are explanatory in the same way" (191). To see whether this is correct, one must ask what it is to identify a property, such as fitness, as a disposition, and, second, what does it explain and how does it operate in an explanation?

I briefly outline how the strategy of dispositional analysis was successful in the case of biological fitness. Biological fitness refers to the correspondence between an organism and its environment. The more fit the organism is, the better it survives and reproduces. An often-cited criticism of evolutionary theory is that the principle of natural selection is tautological. Since this principle seems to be intended to be empirical and not definitional, so much the worse for the scientific status of evolutionary theory. The principle of natural selection is characterized in terms of the Spencerian motto of "the survival of the fittest," or, to put it in a more acceptable form, "Those organisms that are more fit will reproduce more successfully." If, however, fitness is defined as actual reproductive success, an operational definition found, for example, in population genetics, then the principle of natural selection becomes the tautology. "Those organisms that have greater actual reproductive success will survive to reproduce more successfully." The propensity interpretation of fitness (Brandon 1978; Brandon and Beatty 1984; Mills and Beatty 1979) argues that this is a mistaken interpretation of fitness. Identical twins in the same environment are equally fit to that environment. Nevertheless, by chance, one may perish before it reproduces while the other does not. Thus, while the actual reproductive success of the twins differs, their fitness does not. If fitness is construed not as actual reproductive success but as the propensity to have a certain reproductive success, or the expected reproductive success, then the alleged tautology is transformed into an empirical explanatory principle. The twins have identical dispositions, but only one of them actualizes or manifests its potential.

Dispositional properties are such that they can inhere in an object without ever being expressed. A lump of sugar is soluble even if it never dissolves. On the classic reductionist account, a disposition is identified with the subjunctive conditional describing the behavior of an object, the manifestation of the disposition, when certain antecedent conditions are realized. On this view, dispositions are not explanatory properties, but rather shorthand summaries of phenomenological behavior. Clearly, in the arguments to which Bigelow and Pargetter appeal, a nonexplanatory interpretation of dispositions would fail to grant biological fitness the appropriate status to export to biological function. Thus, a realist interpretation of dispositions is taken to be assumed by the propensity interpretation of fitness.

Realists argue that dispositions can be explanatory, but only when associated with a causal basis.[26] The basis can be a physical structure such that having it while being in the antecedent circumstances will cause the manifestation described by the subjective conditional. For example, solubility is a disposition characterized by the conditional "if this object should be placed in water, then it will dissolve," where the manifestation, dissolving, is caused ceteris paribus by meeting the antecedent circumstances of being placed in water and having the appropriate molecular structure. When we ask why this lump of sugar dissolved, we might explain it by its disposition, that is, because it is soluble. Or we might explain it by its causal basis, because it has molecular structure M. The first appeals to the general disposition, which abstracts away from the particular material of the causal basis; the second appeals to that basis.[27]

---

[26] Armstrong 1969, Prior 1985, and Prior, Pargetter, and Jackson (1982) defend the necessity of a causal basis, but seem to allow for the disposition, that is, conditional behavior, to supervene over a variety of different causal bases. Sober, on the other hand, defends the reality of dispositions as similarly independent of the behavioral subjunctive conditional description, but requires something stronger, namely, that "[a] scientifically respectable dispositional property must be a univocal characteristic that underlies all the instances in which the subjective conditional displays itself" (Sober 1984a: 47).

[27] General issues concerning the explanatory status of dispositions arise in the particular case of biological fitness. Namely, here has been discussion about how "fitness" is to be understood dispositionally, namely, whether (i) dispositional language itself is nonexplanatory and hence that fitness be taken as an undefined primitive (Rosenberg 1982, 1985); (ii) "fitness" as a dispositional term merely marks a temporary stage in scientific development, and should be discarded since biology has made sufficient advances to fill out the placeholder for which it was used

It has been argued that biological fitness should be construed as a dispositional property (Brandon 1980; Brandon and Beatty 1984; Mills and Beatty 1979). Hence it would be characterized by the subjunctive "if the organism (or organism type) should be in environment E, then it will leave N offspring." However, fitness is a probabilistic disposition rather than an all-or-nothing disposition such as solubility. Fitness is such that having a particular causal basis, for example, dark wing color for moths or larger size in male primates, when in a given set of environmental circumstances confers probabilities onto the individual or type to display specific levels of reproductive success. So in E, organism O may have 6, 12, or 18 offspring with a probability assigned to each item in the distribution. This can be expressed as the disposition to have an expected number of offspring.[28]

But what is it that fitness is intended to explain when it is interpreted as a dispositional property? Differential fitness between organisms or organism types is supposed to explain why one individual or type of individual survives and reproduces more successfully than another individual or type of individual. Why did primate A reproduce more offspring than primate B? Answer, because primate A was more fit than primate B.

How does this work as an explanation? I believe there are two answers to this question. The first appeals to overall fitness specified only as the disposition or propensity to leave N offspring. In that case, this disposition is explanatory in the same way that a general law is explanatory.[29] It displays Darwin's original insight that what he was seeing in the Galapagos finches and the English pigeons was evidence of the same causal process, namely, that natural selection operates by allowing any advantageous character to be "preserved" through time via differential reproduction of its bearers, given the heritability of the character. Thus, the disposition to leave more offspring will describe a property that is often important to evolutionary change. On the global or abstract level, the disposition to leave N offspring can explain why the change from dark-winged to light-winged in populations of peppered moths is like the change from thin beaks to thick beaks in certain Galapagos finches.

On a more local or concrete level, the disposition to leave N offspring explains why dark-winged moths are more reproductively successful than

---

(Waters 1986); or (iii) the dispositional character of "fitness" is still required to be open-ended (Nordmann 1990; Resnik 1988). In the argument that follows, I defend an explanatory role for the abstract placeholder and thus in this respect share the views of Nordmann and Resnik.

[28] This is the view presented by Mills and Beatty 1979: 270–275.

[29] See Cartwright 1986 for a similar argument regarding teleological explanations in general, and Kitcher 1981 on the unifying role of schematic explanations.

light-winged moths by appealing to the causal basis that is presupposed by the disposition. The local explanation appeals to the causal basis – in this case, dark wing coloration – which in conjunction with environmental conditions allows the manifestation of a reproductive success of N.

In either case the dispositional property, fitness N, explains actual reproductive success. By construing fitness as a disposition, we see what fitness is both at the concrete level, by attaching it to a specific causal basis in a specific population, and at the abstract level, by seeing how whatever the particulars of the causal basis, whether they be dark wings or thick beaks, having such a disposition can have consequences for the processes of natural selection and perhaps evolution.[30]

Thus, the fitness of an organism, being a disposition or propensity to leave an expected number of offspring, can explain why the organism realizes a certain reproductive success, just as the disposition of solubility can explain why a bit of matter dissolves. Fitness is *forward-looking* by referring to a future reproductive success. Indeed, fitness is essentially dispositional in that it cannot be replaced by a list of concrete properties such as large size, dark color, or fleetness. If fitness is identified with actual, realized reproductive success, however, then it cannot explain why one organism is more reproductively successful than another.

But what now of functions? Do they work the same way?

The Problem Shift

As outlined above, the classic philosophical puzzle regarding functions is how could something that is a consequence of the presence of a trait explain why the trait itself is present? I have argued that the etiological theory solves this puzzle by showing how having the functional consequence in the past was responsible for the current presence of the item in question. Both selection *for* the functional consequence and replication of the structure *with* that consequence because of selection are required to warrant function ascription on the etiological account.

Bigelow and Pargetter objected to the causal history strategy because it is too *backward-looking*. They argued by analogy with biological fitness that we should understand functions as dispositions, that is, occurrent properties that have a certain consequence under specified conditions. For Bigelow and Pargetter, "something has a function just when it confers a survival-enhancing

---

[30] Ayala (1970) had something like this in mind when he spoke of two levels of teleology in organisms, the proximate end of features of organisms, as well as the ultimate goal of contribution to increase in reproductive success.

propensity on a creature that possesses it" (1987: 192). But what do these sorts of properties, dispositional functions, allow us to explain? Bigelow and Pargetter say that "functions will be explanatory of survival, just as dispositions are explanatory of their manifestations; for they will explain survival by positing the existence of a character or structure in virtue of which the creature has a propensity to survive" (193).

Applied to the case of sexual dimorphism discussed above, we get the following claims. Let us say that the function of large body size in male primates in a specified environment is to contribute to success in male-to-male competition for mates. What does that mean? On the etiological view, it is clear that identifying the function explains why the males are larger by appealing to the causal history of natural selection and evolution that took the differential effect on male-to-male competition as a cause for differential troop acquisition and hence differential reproductive success. Given the heritability of the trait, this selection process would propagate larger body size over smaller size through a series of generations. On the dispositional view, identifying the function as a contribution to success in conflict is to say that this is a survival-enhancing propensity (although the conflicts usually don't have such direct effects on physical survival of the winners and losers, but rather on their acquisition to mates in this case) that accrues to the trait in question. Hence when the right conditions obtain – that is, two males meet in competition – the function will be manifested in actual domination and troop acquisition and thereby enhance the survival of the larger male. What the function explains on the dispositional view is why a large male primate reproduces more successfully than a small one. What the function explains on the etiological view is why male primates are large.

Another way to put this distinction is as follows: The dispositional account tells us how having a trait with a function, a disposition to have a certain consequence, contributes to the survival and reproductive success of individuals with that trait, that is, how they fare in the struggle for existence that constitutes natural selection. In contrast, the etiological view tells us how having a trait with a function, a consequence that has played a certain role in its causal history, contributes to the presence of the trait, that is, how the trait has evolved by means of natural selection.

The application of a dispositional analysis shows how fitness is explanatory by showing how having a certain fitness could explain its manifestation, a certain reproductive success. The same strategy applied to functions shifts the very question we set out to answer. On the dispositional view, functions similarly explain why the manifestation – that is, survival and reproduction – obtains, by appealing to a survival-enhancing propensity. What it cannot explain is why the trait itself, with its survival-enhancing or reproduction-enhancing propensity, is present in the population, why it has evolved.

Compatibility of Etiologies and Dispositions

Although they address different explanatory projects, are the two theories of functions incompatible? One way to decide this issue is to consider what conclusions about the ascription of function are made by the two theories, that is, how each draws the function/accident distinction.

Recall the well-known case of the evolution of dark wing color in *Biston betularia*. H. B. Kettlewell investigated the evolutionary change in populations of the peppered moth in the environment surrounding Manchester, England, from 1848 to 1898. During this time, the dark-winged or melanic form of the moth, a condition produced by mutation at a single locus, increased in frequency from 1 percent to 99 percent. He argued that the change in the environment during a period of industrialization produced a change in selection pressure responsible. That is, the moths rest on trees during the day and are subject to predation by birds. Before 1848, the trees in the area were covered with lichen and the light-winged variety would be virtually invisible to predators, while the dark-winged variety would be visible. With industrialization, the lichen-covered trees became covered with soot and the lichen died, creating a new environmental condition for the moths. In the new situation, the dark-winged variety was camouflaged from predatory birds, while the light-winged variety was visible and suffered greater mortality as a result. Kettlewell (1956) conducted experiments that showed that birds could indeed detect the dark moths on light backgrounds and light moths on dark backgrounds. Thus, one can justify the claim that the function of dark wing color in the peppered moth is to provide camouflage from predatory birds. That is the consequence responsible for the selection via differential mortality and hence reproduction of dark- over light-winged variants, and that selective process issued in greater replication of the gene for melanism and hence the dark phenotype.

Dark wing color in the peppered moth may have consequences of greater heat absorption as well as camouflage. Which consequences qualify as functions? For the etiological theory, it is the actual historical process that constitutes the criterion for function ascription. Whatever things are like in the present environment, it is the consequence that was responsible for the selection and evolution of the trait in the environmental context of its evolution that is the proper function of that trait. If there had been a selective pressure for the ability to absorb heat, say, in an environment of severe fluctuations of temperature or light intensity, then heat absorption would be the function of dark wings; if not, then it is merely an accidental consequence of the trait.

On the dispositional account, what grants the status of function to a consequence is its ability to enhance survival in a forward-looking fashion. But what environment is relevant to ascribing such a function? Clearly, in different environments different consequences of a trait will enhance survival. If one is designing a zoo, for example, one would be interested in the consequences of traits on survival in what is an artificial environment, perhaps one that would never have existed in nature, and hence one that clearly would not duplicate the environment of selection and evolution that is the concern on the etiological project. What is a function, that is, a survival-enhancing propensity in a glass and metal enclosure with only artificial light, may not be a function in a virtually limitless field in the open air occupied by predators and conspecifics. If we are interested in survival in the zoo, functions will be ascribed according to survival enhancement in that environment. The heat absorption consequence of dark wing color may be a function here, if in fact it allows dark moths to survive under artificial light where light moths would perish.

Suppose, however, that the practitioner of the dispositional theory is interested in survival in the environment of evolutionary history, then the dispositional theory will ascribe functions compatible with the functions ascribed on the etiological theory. The dispositional account has, if you like, weaker criteria or interest-dependent criteria for application. When the interests coincide with those of the evolutionary biologist, then function ascription will not differ greatly from the etiological theory. When the interests are different from those of the evolutionary biologist, however, then the two theories may deliver incompatible classifications of functional and nonfunctional consequences for a trait in question.

Conclusion

The classic puzzle regarding functions is, How could something that is the consequence of the presence of a trait explain why the trait is in fact present, since the function is temporally later and so cannot be causal in a straightforward sense? As we have seen, the etiological account solves this puzzle by showing how having the functional consequence in the past was responsible for the current presence of the item in question. Both selection *for* the functional consequence and replication *of* the structure with that consequence because of selection are required to warrant function ascription on the etiological account.

Wesley Salmon (1990) has claimed that Bigelow and Pargetter's account is both a possible improvement on and not very different from the etiological theory as developed by Larry Wright. I believe Salmon's assessment is mistaken. I have argued that the dispositional theory does not offer a competing analysis to the etiological theory of functional explanation and, as such, cannot be taken as an improvement. Contrary to Salmon, the two theories of function are different in that they are designed to analyze the explanatory ability of function claims as answers to two different questions. When the domains of interest of the two explanatory projects coincide, the two solutions will be compatible. The dispositional theory can explain how traits fare in the process of selection, and the etiological theory extends the explanation to specify how the selection process further affects evolutionary replication. The two need not be applied to the same domain of interest, however, as the dispositional account does not require the restriction of interest to a domain of evolutionary environments.

This is not to say that Bigelow and Pargetter's dispositional analysis explains nothing. Rather, scientific explanations are developed to do different things: to be directed at different targets and to give explanations at different levels of abstraction or concreteness. As Cartwright (1986: 203) puts it, "[e]xplanations give answers not only to *why* questions, but also to *what* questions. They say *of* something, what it really is." Applying this view to the explanatory status of biological fitness, we have seen that appeal to differential fitness works both by directing attention to the structural causal basis to answer why this particular individual survived rather than another one and at the abstract level by explaining what sort of character fitness is in terms of the role it might play in long-term evolutionary processes. One must acknowledge that there are many explanatory projects that constitute scientific investigations.[31] Given that, confusion can arise if one fails clearly to identify one's own project or if one conflates distinct projects. The danger is in postulating a problem solution in one enterprise to do the work in another. Bigelow and Pargetter's defense of the dispositional theory of functions as a replacement for the etiological theory commits just that error.

---

[31] See Wouters 1995 for other types of explanatory projects associated with function.

II

Pluralism

5

Laws

The argument of this book is that complexity in nature, particularly in biological nature, has direct implications for our scientific theories, models, and explanations. In other words, nature is complex and so, too, should be our representations of it. One feature of this corresponding complexity of representations is pluralism, in contrast to unity of scientific constructions. For example, not all apparently similar phenotypes will have the *same* explanation. Selection operating at different levels may produce the same outcome; we cannot reduce selection theory to a single unit or level. Multiple causal components operating simultaneously contribute to biological traits such as complex behaviors; we should not expect a single cause, natural selection, to explain all that we observe. And interactions among parts of complex wholes generate order by the dynamics of their interactions; we should not anticipate simple, linear models to account for the structure and behavior of such systems. In short, the multilevel, multicomponent, complex systems that populate the domain of biology are ill suited to a simple, unified picture of scientific theorizing. Pluralism in this domain is not an embarrassment of an immature science, but the mark of a science of complexity.

A notorious example of the mismatch between the expectations of a reductive unificationist and the practices of biologists is the absence of biological laws. Universal, exceptionless laws have been taken to be the hallmark of successful science. Biology does not produce anything that looks like these idols of philosophy of science, and hence its genuine scientific status comes into question. If universal, exceptionless laws are what ground explanations, guarantee prediction, and permit us to intervene successfully in the world, then biology's lack of them is a serious problem indeed. But is this standard account of laws the correct one? In the following sections, I suggest there is another interpretation of laws, a pragmatic one, which renders the knowledge acquired of complex, biological phenomena lawful, explanatory, and predictive.

Introduction

In a recent paper, Beatty (1995) argued for what he calls the *evolutionary contingency thesis (ECT)*.[1] This is the claim that generalizations about the living world are mathematical, physical, or chemical laws or are distinctively biological in that they describe contingent outcomes of evolution. Beatty takes this to imply that there are no genuine biological laws, because "whatever 'laws' are, they are supposed to be more than just contingently true" (46). Brandon (1997) and Sober (1997) endorse the conclusion that insofar as the generalizations of biology are contingent, they fail to be laws, and Beatty (1997) explores further support for ECT from scientific disputes. I agree with the substantive claims of Beatty and Brandon concerning the use of biological generalizations in scientific practice and find no logical error in Sober's formal representation of these generalizations. Nevertheless, I argue that these papers, like most discussions of biology's failure to produce genuine scientific laws, are limited by their shared normative approach to the question. After demonstrating these limitations, I sketch an alternative strategy – a pragmatic approach to laws – and indicate how it provides a more adequate representational framework in which to characterize two important features of scientific practice:

- the variability of types of generalization within the empirical sciences, and
- the *nature* and *degree* of contingency characteristic of biological generalizations.

This section is an attempt to direct the entire discussion away from the question of what we should call a law toward an understanding of how scientific generalizations of various types function in inferences to satisfy the pragmatic goals of science.

How do we decide whether or not biology has laws? There are three strategies for pursuing this question: a normative, a paradigmatic, and a pragmatic approach. The normative approach is the most familiar. To proceed, one begins with a norm or *definition* of lawfulness, and then each candidate

---

[1] This section was originally published as S. D. Mitchell, "Pragmatic Laws," *Philosophy of Science* 64 (1997): S468–S479.

generalization in biology is reviewed to see whether the specified conditions are met. If yes, then there are laws in biology; if no, then there are not laws in biology. The paradigmatic approach begins with a set of *exemplars* of laws (characteristically in physics) and compares these with the generalizations of biology. Again, if a match is found, then biology is deemed lawful. The pragmatic approach focuses on the *role* of laws in science and queries biological generalizations to see whether and to what degree they function in that role.

The Normative Strategy

Beatty (1997), Brandon (1997), and Sober (1997) all acknowledge the legacy of logical empiricism in their questioning the existence of biological laws. In that tradition, laws were initially characterized syntactically as universal generalizations in first-order predicate calculus. The problem of ruling out the merely accidentally true generalizations that share this form forced attempts to further restrict the definition. These intuitions are familiar. That the diameter of a sphere of enriched uranium never exceeds 100 meters instantiates a law of nature. That the diameter of a sphere of gold never exceeds the same length is true, but accidentally so. What is required to draw this distinction is a way to isolate the necessitation of the consequent condition upon the action of the antecedent in order to cover only those generalizations that *could* never, not just *had* never, failed to be true. This feature – what has been called natural necessity – explains why laws, and not accidentally true generalizations, support counterfactual conditionals, can be confirmed by a small number of positive instances, and are projectible. Like much in the logical empiricist tradition, this notion of natural necessity was fashioned from the cloth of logic. Some of the problems with this approach are artifacts of that history. I suggest that insofar as natural necessity closely mirrors logical necessity, it will fail to adequately characterize the empirical relations investigated by science.

Logical necessity is an all-or-nothing affair. Either a statement's truth follows necessarily from the truth of a set of statements, or in virtue of its form, without exception and all times, or it does not. The security of expectation warranted by logical necessity may well be comforting, but that security does not get carried along when expropriating the notion of necessity to the natural world. The truth of a statement that is not logically necessary is contingent on other things. All naturally necessary relations represented in lawlike statements in science are contingent in this logical sense. Hence the distinction between naturally necessary and merely accidental generalizations simply cannot be drawn on the presence or absence of contingency per se. The dichotomous character of the distinction (logically necessary/contingent, naturally necessary/accidental) must be abandoned. It is the *nature* and the *degree* of contingency, and not the fact of contingency, that separates the lawful from the accidental. Making this explicit forces a move out of the dichotomous space inherited from logical definitions into a continuous domain of kinds and degrees of contingency that may be exhibited by scientific generalizations.

A limitation of the normative approach can be seen in the disagreement between Sober and Beatty. While Sober agrees with Beatty that biological generalizations, if contingent, would fail to be laws, he suggests, first, that there are a priori noncontingent biological laws, and second, that Beatty's account of evolutionary contingency entails that for every contingent biological law discovered, there must be a noncontingent law in which it can be embedded. Sober's argument depends on articulating the implicit ceteris paribus conditions antecedent in all scientific laws. The necessitation relation described by a law holds only when the assumed boundary conditions are also met. I argue that this way to represent the problem obscures just those features of contingency that Beatty ascribes to the characteristically biological.

Beatty claims that distinctively biological generalizations, while true, are contingent on a particular historical pathway traversed as a result of evolutionary dynamics. Mendel's law of the 50:50 ratio of gamete segregation is true only because the genes determining that ratio had been selected for in a particular episode in the evolutionary history of life on this planet. If we had, in Gould's words, "run the tape again," it is likely that different genes would have been available through mutation, different traits would have evolved, and hence different generalizations would then be true. Indeed, those historical conditions on which the truth of a generalization is contingent (e.g., those determining the selective advantage of the 50:50 segregation gene) may change in the future, rendering the generalization no longer capable of truly describing the state of nature. Beatty calls this feature "weak contingency." In addition, by "strong contingency" Beatty denotes the fact that from the same set of conditions with the same selection pressures operating, variant functionally equivalent outcomes may be generated. "To say that biological generalizations are evolutionarily contingent is to say that they are not laws of nature – they do not express any natural necessity; they may be true, but nothing in nature necessitates their truth" (Beatty 1995: 52).

Sober represents Beatty's thesis using the following logical formulation:

$$\frac{I \; \rightarrow \; [\text{if } P \text{ then } Q]}{t_0 \qquad t_1 \qquad t_2}$$

Here [if P then Q] is an evolutionarily contingent generalization and I represents the historical conditions on which it depends. Sober argues that in using this representation we can easily see that there really is a noncontingent biological law being invoked: (L): $I \rightarrow$ [if P then Q]. Sober claims that here are two ways a law of the form L can escape contingency, either by being *analytic* or by taking seriously Beatty's causal claim that there was a particular set of conditions in evolutionary history responsible for $P \rightarrow Q$ being true. That the mathematical laws used in biology (such as Hardy-Weinberg equilibrium) are logical truths is neither controversial nor particularly pertinent. That Beatty's very argument for contingency is self-refuting is more serious.

Notice what must be presumed to allow Sober the second interpretation. The assumed complex web of present and absent environmental conditions that rendered the 50:50 gene more fit than other variants, that did not trade off those consequences via other selective pressures, that prevented chance and mutation and migration from overriding the advantage, and so on, is represented simply by I. In addition, the complicated causal process that gave rise to the relation described by [if P then Q] is abstracted to the arrow of the material conditional. In what way does the new law, L, describe a naturally necessary relation between the antecedent, I, and the consequent, [if P then Q]? One could argue that if all the conditions that cause the rule to be true have been identified, then their occurrence must unconditionally necessitate the truth of the rule. If not, then one could counter that the complete causal story was not described. That is, insofar as Beatty's claim that evolutionary conditions and processes caused the generalization to be true, then a noncontingent causal generalization, namely, L, should be capable of describing that. But is it really noncontingent? Suppose L is a physical rather than a biological law. Fourier's law of heat conduction, for example, describes necessary causal relations on the presumption that our universe is in a state of thermal disequilibrium (Carrier 1995). As far as we know this condition was fixed by the distribution of particles in the primordial atom, nevermore to change unless, that is, the universe were to suffer heat death. Indeed, one expects the relation described by Fourier's law to be the only physically possible true relation of conduction because it depends on a very stable and enduring set of I conditions. Since the presumption is that thermal disequilibrium is a standing condition, articulating it explicitly in I is a mere formality. Sober transforms the explicitly contingent biological generalization not into a noncontingent law, but rather into an implicitly contingent physical law. This logical sleight-of-hand obscures, rather than illuminates, the similarities and differences between the evolutionary and physical relations that causally structure our world.

Two forms of complexity are hidden in Sober's representation. First, the set of conditions, I, in the biological story consists of a complex and unstable conjunction of conditions. In addition, the causal story that lurks in the material conditional is also complex in the sense of being nonlinear. That means that very minor variations in the complex set I could lead to very dramatic differences in the consequent represented by [if P then Q]. Even granting that the conditions specified by I cause the relation described by the consequent rule does not preclude variant outcomes from being determined. The simplicity of $I \rightarrow$ [if P then Q] conceals what is distinctively biological, namely, both the complexity of the conditions upon which the law is contingent and the complexity of the nature of the dependence. In contrast, while a physical law, like Fourier's law, can also be represented as [if P then Q] and the historically contingent arrangement of the primordial atom be identified as I, the similarity ends there. The I conditions are stable, in that arguably they were fixed in the first three minutes of the birth of the universe and are extremely unlikely to change. In addition, in the absence of thermal equilibrium, Fourier's law arguably describes the unique relation of conduction true for our universe.

To summarize, Sober is correct that one may reformulate the weakly and strongly contingent biological generalizations that Beatty documents in a simple, logical representation that looks like a physical law. But what is lost is just the characteristic complexity of biological conditions and causes. It is not *that* biological generalizations are contingent, but rather *how* they are contingent that is significant. Importantly, practices of biologists differ from those of physical scientists in ways that correspond to these differences.

With respect to this latter point, Beatty attempts to read the nature of biological generalizations backward in an abductive, or transcendental, inference from the form of the relative significance debates in which biologists are engaged. What, we may ask, must the biological world be like to make sense of such a practice? I think this strategy is powerful, and I have engaged in a similar type of argument to explain what I have called the "fact of pluralism" of explanatory models in biology (Mitchell 1997). The conclusion Beatty draws is that laws in biology are not universal and exceptionless, since the debates presuppose generalizations of limited scope. Nevertheless, Beatty's analysis of relative significance debates suffers from two ambiguities.

First, the debate structure of biologists' discourse is surely a function not just of the ontological conditions of the biological world, but also of the biologists' understanding of scientific laws. If they have been properly schooled with the Popperians, as many of them have, then they should be looking for bold, universal, exceptionless laws, and their failure to find them will look like a failure to find laws at all. There still may be laws in biology, but they will not be recognizable as such by those blinded by a limited normative definition. So, while Beatty's approach has the potential to identify a mismatch between a particular *view* of scientific laws and the practices of biologists, it does not yet address whether or not there might be "laws" otherwise construed in biology.

Second, there is a conflation of two distinct problems in Beatty's claim that "[t]he relative importance or significance of a theory within its intended domain is roughly the proportion of phenomena within the domain that the theory correctly describes" (1997). Biological explanations invoke multiple models for two reasons:

- a multiplicity of causal factors interact in generating complex phenomena, and
- different causal factors are restricted to only partially overlapping spatiotemporal domains.

Consider first Beatty's discussion of the lac operon theory. Though it was originally promoted to explain gene regulation in all organisms, it was later discovered not to be adequate to the task. Additional mechanisms (negative repression, positive induction and repression, and attenuation theory) were needed to describe the spatiotemporal diversity of systems from colon bacillus to the elephant. No one mechanism or combination of mechanisms was universal. In contrast, the debate between selectionist and neutralist theories of microevolution is not always about whether just selection or just neutralism operates most frequently in determining evolutionary outcomes across the domain, but also concerns the relative contribution of each causal factor in generating a specific complex outcome. I have argued elsewhere for the theoretical pluralism of idealized models and the necessary integration of explanation required to account for multifactor complexity (see Mitchell 1992, 1997; Mitchell et al. 1997). Here, I just want to point out that these ways of failing to be exceptionless laws are different. While both support theoretical pluralism, they entail different scientific practices in generating acceptable explanations. Within the confines of the normative approach, these problems are not prima facie distinguishable. Rather, they are classified identically as failures to be universal, exceptionless laws.

The normative approach's set of necessary and sufficient conditions that must be met by generalizations to qualify as laws provides a limited conceptual space in which to explore the important differences among biological generalizations and between them and those of physics. A law is necessary, or it is not. Sober's logical representation also fails in this regard. It is like trying to describe the differences between Beatty, Sober, and Brandon by first representing them as stick figures. More must be done to enrich the contingency/noncontingency distinction in order to adequately describe the varied types of generalizations explored in the many sciences. That biological generalizations, [if P then Q], are contingent on conditions I does not distinguish them from physics or chemistry, for these, too, will reference some more basic conditions that are true of our world, are not logically necessary, and on which the truth of the laws within those sciences also depends. It is only by attending to the *nature* and *degree* of contingency that a proper understanding of scientific generalizations can be developed.

The Paradigmatic Strategy

Let me turn briefly to a paradigmatic approach to answering the question of whether or not the generalizations of biology are laws. Here one engages in a primarily descriptive project that begins with identifying exemplars of laws in physical science, such as Newton's laws of motion or the ideal gas law, and proceeds to examine the candidate generalizations in biology to see whether they are similar to the paradigmatic laws. This is the strategy Carrier (1995) adopted in his critique of Beatty's ECT. Carrier claims that scientific laws should support counterfactuals, but, pace Beatty, he concludes that biological generalizations are lawlike. What he provides is a series of arguments by analogy, detailing the similarity between biological generalizations and their analogs in physics. Thus, insofar as the paradigmatic physical laws are laws, and biological generalizations are like them, then biology, too, has laws. He shows that all physical laws invoke boundary conditions, and some physical laws, like those for high- and low-level superconductivity, have restricted domains of applicability. In addition, he appeals to idealized formulations, such as Newton's first law of motion, as evidence of the exception-prone character of some members of the exemplar set.

Rather than explore this strategy in greater detail, I wish only to point out two problems in pursuing it. The first is a feature of lumping together all the exemplar laws of physics undifferentiated with the status of law. Specifically, Carrier seems at times to confuse the dependence of the truth of the consequent of the law on the conditions of the antecedent being true (as in Newton's law of inertia) with the relations described by the law themselves being contingent on historically specific events (as in the superconductivity laws). This confuses the contingent relation described in the law (Q is contingent on P) with the dependence of that relation on other conditions ($I \rightarrow$ [if P then Q]). Second, and more important, while taking physical laws as paradigmatic and comparing them with biological generalizations is a useful enterprise, it leaves open the philosophical question of what a law of nature is. Biology on this account is no worse off than physics or chemistry, and as long as our intuitions about what counts as a law are secure in the exemplar domain, then our evaluation of biological generalizations will follow. However, it fails to address the underlying question of what it is about the cases we identify as exemplar laws that makes them laws in the first place. Carrier is certainly aware of this problem, suggesting that given his arguments, in the end, we could just as well say that there are no laws anywhere in the sciences. Indeed, as I have suggested, there are significant differences both within the set of laws of a given science, such as physics or biology, and between them. As this paper argues, a richer conceptual framework allows a detailed account of these similarities and differences to be explored.

The Pragmatic Strategy

Taking a pragmatic approach to scientific laws replaces a definitional norm and multiple exemplars with an account of the *use* of scientific laws. How do they function in experiment, in explanation, in education, or in engineering? The features of generalizations that perform in these roles can be determined, and one proceeds to see whether and how the generalizations in biology function as laws. The result is a framework for representing the multiple types of generalization found in the various sciences. Notice that rather than a dichotomous space defined by the normative approach, or the unsystematized space of the paradigmatic approach, this view supports a multidimensional frame in which to view these varied qualities of scientific generalizations. This enhances the investigation of multiple lines of relationship among generalizations both within a scientific domain such as physics or biology and between these domains. Brandon's (1997) discussion of types of experimentation can be seen as a contribution to this philosophical enterprise.

An important insight in Brandon's discussion is the recognition that the multiplicity of practices in biology are driven by both the ontology of the biological world and the special interests of the scientific community. He shows that the reason biology, compared with physics, engages in less manipulative and more parameter-setting experimental practices is a function of both the nonprojectible, contingent relations investigated and the history of the epistemic community. Nevertheless, Brandon holds to the normative definition that, to be laws, generalizations

1. have nomic or natural necessity;
2. are used essentially in scientific explanation; and
3. receive confirmation from (a small number of) their positive instances (Brandon 1997).

Brandon evokes the tension between the acknowledged explanatory character of biological generalizations and their failure to meet the stringent conditions of exceptionless universality. In his words: "the contingent regularities of biology have (a limited range of) nomic necessity and have (a limited range of) explanatory power, but lack . . . unlimited projectibility" (Brandon 1997). Brandon defends the ability of biological generalizations to explain phenomena and thereby function to fulfill one of the clear goals of scientific practice while recognizing their failure to satisfy the specified norm of lawfulness. For him, the tension is resolved by surrendering their lawlike status. Biological generalizations are projectible and ground explanations but only within a less than global range. How limited the range is varies and must be discovered empirically.

While I agree with Brandon's description of the way biological generalizations are used in explanation, I disagree that the best way to acknowledge the special character of biology is to remain loyal to the limited dichotomous conceptual framework of the normative approach to laws. By doing so, one privileges a form of generalization that occurs only rarely, if at all, even in physics. Brandon is led to this conclusion by rejecting what he takes to be the only alternative, namely, merely extending law status by global edict to broader categories of generalizations in order to cover the evolutionarily contingent ones. I also reject this possibility, as it, too, would fail to provide resources for identifying differences in complexity evidenced by multiple component causes, nonlinear causal relations, and unstable conditionalizations. The pragmatic approach offers an alternative.

The function of scientific generalizations is to provide reliable expectations of the occurrence of events and patterns of properties. The tools we design and use for this are true generalizations that describe the actual structures that persist in the natural world. The ideal situation would be, of course, if we could always detach the generalizations gleaned from specific investigations from their supporting evidence, carry these laws to all regions of space-time, and be ensured of their applicability. Such generalizations would be universal and exceptionless. But some causal structures – in particular, those studied by biology – are not global. Thus, the generalizations describing them cannot be completely detached from their supporting evidence. Nevertheless, we can and do develop appropriate expectations without the aid of general-purpose tools – laws that govern all time and space without exception or failure. To know when to rely on a generalization we need to know when it will apply, and this can be decided only from knowing under what specific conditions it has applied before. To use Sober's representation, the conditions I upon which [if P then Q] is contingent may be located on a continuum of stability. In addition, the nature of the dependence relation, $\rightarrow$, reflects a continuum of strength, including probabilities and multiple determinant outcomes. Life, it turns out, is not as simple as we might have hoped. Our representations of it will be correspondingly complex.

In addition to the ontological parameters, there are other pragmatic aims for which we use generalizations. Scientific representations can be evaluated for their usefulness in virtue of

- *Degree of accuracy* attuned to specified goals of intervention. The eradication of insect pests may require assessing a relatively crude lawlike relation between a chemical, say, and the death of the insect, while increasing the fecundity of other insects, such as bees, may require describing more detailed mechanisms and relations.
- *Level of ontology*. Generalizations about populations may describe structural relations between trait groups (such as large and small size on calling frequencies in crickets) or functional groups (such as predators and prey). Some relations described appear only at a given level and not above or below.
- *Simplicity*. We use generalizations ranging from rules of thumb such as Ptolemaic astronomical "laws" to navigate to ideal gas laws that yield approximations within engineering tolerances.
- *Cognitive manageability*. Prior to the development of high-speed computation, mathematical equations were restricted to solvable linear formulations.

The contingency of generalizations in biology or other sciences does not preclude their functioning as "laws," or generalizations that ground and inform expectations in a variety of contexts. They are usable, pragmatic laws when they provide a particular expectation (the scope of domains to which we can export an empirically discovered relation) and specify the degree of strength of that expectation (in terms of probability or complexity). Scope and strength are dimensions that can be used to compare generalizations within physics or biology, as well as between them. In the multidimensional space defined by the multiple aims of scientific practice, including the ontological parameters as well as accuracy, simplicity, ontological specificity, and manageability, it may well turn out that all or most of the generalizations of physics occupy a region distinct from the region occupied by generalizations of biology. The conditions on which physical laws are contingent may be more stable through space and time than the contingent relations described in biological laws. The strength of the determination can also vary from low probability relations to full-fledged determinism, from unique to multiple outcomes. Indeed, the causal contribution of particular features may vary in their sensitivity to environmental conditions, including the presence or absence of other causal factors. While I have only sketched the parameters by which generalizations may be compared, it is clear that such a conceptual framework has the resources to display the multiple relationships that exist among and between generalizations in the sciences.

Rather than bemoan the failure of biological generalizations to live up to the normative definition of exceptionless universality, the pragmatic approach suggests a different philosophical project. To understand the multiple relations among scientific generalizations, one must first explore the parameters that make generalizations useful in grounding expectation in a variety of contexts.

5.2. DIMENSIONS OF SCIENTIFIC LAW

Introduction

Biological knowledge does not appear to fit the image of science that philosophers have developed.[2] In particular, it has long been argued that biology has no laws (Beatty 1995; Smart 1968). Yet biologists speak of "laws" in their writings. One of Mendel's "laws" claims that with respect to each pair of alleles at a locus on the chromosome of a sexual organism, 50 percent of the organism's gametes will carry one representative of that pair, and 50 percent will carry the other representative of the pair. Recently, a number of biological scaling "laws" have been discovered. These include Kleiber's law that metabolism increases in proportion to body mass raised to the 3/4 power, and the scaling law that respiratory rate is inversely proportional to body mass raised to the 1/4 power (Pool 1997; West et al. 1997).

[2] The research for this section was supported by National Science Foundation, Science and Technology Studies Program, grant no. 9710615. An earlier version was read at the American Philosophical Association, Eastern Division Meetings, December 1998. This section was originally published as S. D. Mitchell, "Dimensions of Scientific Law," *Philosophy of Science* 67 (2000): 242–265.

Why do some philosophers fail to count these results of biological investigation as laws? How are they different from Proust's law of definite proportion, or Galileo's law of free fall or the conservation of mass-energy law? Those who argue that there are no laws in biology point to the historical contingency of biological structures and the particularity of the referents in biological generalizations as grounds for excluding the law designation. In considering the problem of the existence of biological laws I was led to a general reflection on laws in science. My conclusion is that we need to think about scientific laws in a very different way: to recognize a multidimensional framework in which knowledge claims may be located and to use this more complex framework to explore the variety of epistemic practices that constitute science. In this section, I argue that dichotomous oppositions such as "law versus accident" and "necessity versus contingency" produce an impoverished conceptual framework that obscures much interesting variation in both the types of causal structures studied by the sciences and the types of representations used by scientists.

I take this project to be similar in spirit to Carnap's analysis of the acceptance of different linguistic forms within science. He concludes his investigation of the relative worth of using thing language, abstract language, or not speaking at all:

> The acceptance or rejection of abstract linguistic forms, just as the acceptance or rejection of any other linguistic forms in any branch of science, will finally be decided by their efficiency as instruments, the ratio of the results achieved to the amount and complexity of the efforts required. To decree dogmatic prohibitions of certain linguistic forms instead of testing them by their success or failure in practical use is worse than futile; it is positively harmful because it may obstruct scientific progress. The history of science shows examples of such prohibitions based on prejudices deriving from religious, mythological, metaphysical, or other irrational sources, which slowed up the developments for shorter or longer periods of time. Let us learn from the lessons of history. Let us grant to those who work in any special field of investigation the freedom to use any form of expression which seems useful to them; the work in the field will sooner or later lead to the elimination of those forms which have no useful function. Let us be cautious in making assertions and critical in examining them, but tolerant in permitting linguistic forms. (Carnap 1950)

I endorse both Carnap's pragmatic standard and his plea for toleration. However, I believe it should be applied not only to linguistic expressions within science, but also to philosophical expressions about science.

In addressing generalizations used in science, should we talk exclusively in terms of laws and accidents? Should we talk solely in terms of necessity and contingency? Should we represent the type of generality of scientific knowledge only in terms of the universal operator in first-order predicate calculus? Representational forms and particular representations are simultaneously illuminating and limiting. They cannot perfectly represent their objects because they do not display *all* the features of the thing represented. Therefore, they must be judged, at least in part, in terms of their usefulness. In defending a multidimensional account of scientific knowledge, I expose limitations of traditional philosophical analyses and representations of knowledge of causal structures in nature in the hopes of showing how a different sort of enterprise promises to be better for understanding the diversity of scientific practices.

In examining the question of the nature of scientific laws and the existence or nonexistence of laws in biology, it is useful to ask first what options are available for approaching the problem. Before we can decide whether biology has laws or whether any claim of knowledge about the world is a law, we need to be clear about what a law is and what the candidate claims look like.

Normative and Pragmatic Strategies

In Mitchell (1997) I suggested three strategies for investigating the question of the existence of laws in biology: a normative, a paradigmatic, and a pragmatic approach:

> The normative approach is the most familiar. To proceed, one begins with a norm or *definition* of lawfulness and then each candidate generalization in biology is reviewed to see if the specified conditions are met. If yes, then there are laws in biology, if not, then there are no laws in biology. The paradigmatic approach begins with a set of *exemplars* of laws (characteristically in physics) and compares these to the generalizations of biology. Again, if a match is found, then biology is deemed lawful. The pragmatic approach focuses on the *role* of laws in science, and queries biological generalizations to see whether and to what degree they function in that role. (Mitchell 1997: S469)

In this section I further develop those ideas. I elaborate the normative approach, articulating both its features and its foibles. I then argue in favor of a pragmatic approach that focuses on the function of laws and holds that there are a variety of forms of scientific claims that provide us with usable knowledge. I show that the pragmatic strategy leads us to develop a multidimensional framework of features that characterize useful scientific generalizations. I argue that only by substituting this analytical framework for the standard dichotomous law/accident talk do we adequately represent the complexities of good scientific practice.

Traditional Normative Approaches

Considerations of what counts as a law of nature and corresponding implications for the variety of claims generated in the many sciences continues to be dominated by normative approaches. That is, one begins with a *definition* of lawfulness that constitutes the standard by which scientific claims are judged. Among the normativists, the largest divide is between empiricists (of naive or sophisticated sorts) who claim that laws are just true generalizations describing the regular events occurring in nature and necessitarians who claim that laws explain why those events, and not others, obtain. For empiricists, patterns of events constitute laws, while for necessitarians it is the laws that "govern" which events occur (Dretske 1977; Earman 1984). In both cases, the identifying characteristics for what qualifies as a *law* are some notion of generality or universality and some notion of necessity. After all, we use laws to tell us about what happens outside the confines of our finite experience. As Richard Feynman (1995: 164) puts it: "[s]cience is only useful if it tells you about some experiment that has not been done; it is no good if it only tells you what just went on. It is necessary to extend the ideas beyond where they have been tested." Generality has been standardly symbolized by universality in the logical sense, that is, the kind of generality that can be represented as (x) in a statement of law $(x)(Px \rightarrow Qx)$. Necessity is often identified with an ability to "support" counterfactual claims. Knowledge of laws is meant to allow us to predict and explain particular events and hence successfully intervene in our world.

For empiricists this means knowing what the world of events is like, what follows what. Laws are the best summary of those facts. For necessitarians, laws express relations that explain the facts. Necessitarian laws prescribe, not just describe, the events that occur in our world. To accomplish that, necessitarians require knowledge of laws to be more than just a record of what is true, but rather a description of what must be true. Thus, universality and truth, for them, are insufficient. On the basis of universal truths, we could predict what would occur for all time, but there is still a worry about explanation. Some universal truths are taken to be merely accidental and thus incapable of explaining why one event occurred by failing to preclude other possible events that could have occurred in its stead. To have an explanation, it is suggested, one needs to have knowledge of what is possible and not possible. Thus, the necessitarian argument goes, there is more to laws than universal factual truth; there is some form of natural necessity. It is this feature that permits laws (and not merely accidentally true claims about the universe) to "support" counterfactuals and thereby explain particular occurrences in the world.

The problem of accidental truths arises sharply for empiricists in the Humean tradition who find no warrant from experience for necessity above and beyond the warrant for universal truth. They nevertheless want to distinguish laws from other sorts of true claims and appeal to the systematic connections between scientific statements to judge between the lawful and the lawless truths. Thus, those universal truths that occupy a central place in our systematic explanations of the world (or are included as axioms in the set of claims from which we can derive true statements about the world) are deemed laws. "All spheres of gold found naturally on the earth have a diameter of less than 100 meters" is to be distinguished from "All spheres of uranium found naturally on the earth have a diameter of less than 100 meters." Both are true and may be true for all time. But the intuition is that while the latter truth is lawful, the former is accidental. It is the perceived failure of empiricist accounts to *explain* regularities that led necessitarians to opt for a richer ontological picture that locates a law's capacity for explanation in relations among universals, inherent propensities or powers, or patterns of facts among realistically interpreted possible worlds.

A range of views can be found in the normativist camp. My concern here is not with the details that differentiate them, but rather with what the general strategy shares. There is general agreement that laws allow us to explain, predict, and successfully intervene in the world. The features that are supposed to allow them to accomplish these functions are

1. Logical contingency (have empirical content)
2. Universality (cover all space and time)
3. Truth (are exceptionless)
4. Natural necessity (are not accidental)

How do these criteria get interpreted when scrutinizing knowledge claims to determine their lawful or lawless status? Traditionally, philosophers represent scientific claims that appear in either natural language or mathematical formula in some formal logic. Facts are translated into propositional claims, and laws are rendered as universal quantified conditionals (or some properly modalized version of such). $(x)(Px \rightarrow Qx)$ is the familiar reflection of a scientific law in this schema. The functions of laws, that is, explanation and prediction, are then rendered as deductive (or sometimes inductive) patterns of inference from the suitably formalized law statements to suitably rendered fact statements.

The features of laws as they are traditionally understood and the standard ways of representing them has blinded us to important features of scientific knowledge. While the normativist approach has successfully explicated the strongest versions of knowledge claims that can perform the required functions, "There are more things in heaven and earth, Horatio, than are dreamt of in your philosophy" (Shakespeare, *Hamlet*, first quarto, lines 607–608). Part of the problem is a result of the Boolean character of the representational framework of standard logic. Statements are either true or false, and the truth of a statement either follows necessarily from the truth of some other statements (or in virtue of its form) or does not, that is, is only contingently true. The insights that scientists acquire about the causal structure of our world may be deformed by being squeezed into Boolean garb. The problems can be seen by considering each of the traditional defining characteristics in turn.

Scientific laws are empirical truths. Their logical contingency will be reflected, in part, by the logical structure of law statement used to represent them. Thus, $(x)(Px \rightarrow Qx)$ is an acceptable form for a law, while $(x)(Px \rightarrow Px)$ is not. How we accurately represent the discoveries of science is open to interpretation. Are F and ma, in Newton's Second Law of Motion, intersubstitutable equivalents? Mach (1883) took this law to be a definition and hence to have no empirical content. More recently, philosophers have suggested that we treat certain parts of the set of claims that constitute a research program or paradigm as if they were unfalsifiable, thereby rendering them methodologically analytic (Kuhn 1962; Lakatos 1970). Indeed, there are well-worn worries about drawing a clear distinction between analytic and synthetic statements. Thus, the feature of contingency characteristic of laws is not uncontroversially displayed by a representation in first-order predicate logic.

The second characteristic, universality, is standardly represented by the universal quantifier (x) in $(x)(Px \rightarrow Qx)$. The scope of the quantifier is taken to be all space and all time. Hempel's (1945) solution to the "paradoxes of confirmation" problem that vexed this representation of laws makes clear that the intended scope is this broad. Laws are about our world for all time, and hence all occurrences contribute to the confirmation or disconfirmation of a law. On this interpretation it is taken to be ad hoc to restrict the scope of observations taken as relevant to confirmation. Once the scope of the law is understood as universal in this broadest sense, then it is clear that the truth of the law will permit no exceptions. That is, any point in space-time that is described as Pa and $\sim Qa$ excludes the purported law statement from qualifying as a law.

Table 5.1. *Natural Necessity Mirrors Logical Necessity*

Logically necessary	Logically contingent	
	Nomically necessary	vs.	Nomically contingent

The necessitarians emphasize that there is more to laws than universal, contingent truth. Natural necessity is also the mark of the lawful. It is this feature that is supposed to distinguish between so-called accidentally true generalizations and law-like ones, account for the explanatory power of laws, and permit laws to "support" counterfactuals. A problem derives from thinking about natural necessity as isomorphic to logical necessity. Logical necessity carries the strongest possible warrant from truth of premises to truth of conclusion: The conclusion could not be false. A similar, though not identical, kind of warrant is desired to carry one from occurrence of cause to occurrence of effect in the expression of laws about the natural world; the effect could not be otherwise.

Modeling natural necessity on logical necessity carries with it the presumption that the latter, like the former, is an all or nothing property (see Table 5.1). Logically, a statement is either necessary or contingent. So, nomologically a relation between two events in the world is taken to be either necessary or contingent (i.e., accidental). Because the received view has been wedded to representing epistemological relations (such as explanation, prediction, confirmation) and causal relations in first-order predicate logic, it has allowed a reification of the features of the representational apparatus to be imposed on the thing represented. The dichotomous character of logical truth/falsity and necessity/contingency is mirrored in the empirical truth/falsity and nomological necessity/contingency relations.

Boolean representations are taken to reflect causal relations characteristic of our world. This leaves no place, except the vast category of nonlaws, in which to locate a generalization that describes a strong causal relation between events yet fails to exhibit the strongest conditions of nomological connection. Mendel's law of 50:50 segregation pertains to contingently evolved organisms and, even so, has exceptions among those. Thus, on the traditional account, it fails to satisfy the strong warrant attached to necessary laws. The result of judging biological generalizations by the normativist definition of a law is the conclusion that biology has no laws.

A question may be raised at this point as to what the philosophical enterprise of providing an account of laws of nature aims to accomplish. I believe we should begin with what science has discovered about our world that allows us to explain, predict, and successfully intervene. It is clear that scientists, at least sometimes, use the language of laws to capture the causal patterns detected in the results of observations and experimental setups they investigate. In general, what is required for usable knowledge is some claim that one can detach from the particulars of a given observational or experimental situation and export to other contexts (as Feynman clearly states in the quote above). What kinds of information may be used for this purpose? One can attempt to describe the best case, the ideal that, if acquired, would be applicable to all contexts outside the evidentiary ones. It seems to me that this is the type of law that philosophers have attempted to describe. The normativist law is universal, exceptionless, and necessary, and hence is guaranteed to apply everywhere and for all time. This type of claim will certainly function to predict, explain, and allow us to successfully intervene. Yet when one looks to the actual products of scientific practice, one is hard pressed to find examples that fit that ideal image. Rather, what one does find in scientific papers is a range and variety of models, explanations, and theories that provide us with the tools for intervening in our world. Some scientific laws fail to exhibit the ideal properties of philosophical conceptions of law. The mismatch between the object of philosophical theories of law and the products of scientific practices is implied by Mayr's report of the impact of the debate on the nonexistence of laws in biology. He says that "biologists have paid virtually no attention to the argument, implying that this question is of little relevance for the working biologist" (1982: 32).

The working biologist or chemist or social scientist makes do with knowledge claims that fall short of the philosopher's ideal. The appropriate response, I argue, is not to impugn biology, chemistry, and the social sciences for failing to deliver the philosophically valued goods. Rather, this "failure" invites the philosopher to explore just how it is that we manage to explain, predict, and intervene on the basis of these "lesser" variants of lawful relations. How universal, exceptionless, necessarily true generalizations explain, predict, and allow successful intervention is a relatively simple matter compared with how "lesser" variants *actually used in these sciences* manage to perform those same functions. The normativist view of laws and the standard representation of them permits the application of general knowledge of laws to particular events (explanation, prediction, or intervention) by means of instantiation. If $(x)(Px \rightarrow Qx)$ is the law, and here we have or will have a case of Pa, then we know to expect that it will be followed without exception by a case of Qa. But what if there are exceptions? What if the "law" applies much of the time but not all of the time? We can use probability to represent expectations in these cases. But often we don't have information about frequencies, and sometimes we do have information about the kinds of conditioning factors that make it more or less likely for the general relation to hold in particular cases. Making an attempt to explicate the way in which causal knowledge of this sort is usable is more profitable than to just relegate those claims to the heap of "accidents." That there is variation among knowledge claims in science is obvious. An understanding of the significance of those variations may well be lost in adopting the ideal image of a law prior to investigating actual scientific claims.

Biological Laws and the Continuum of Contingency

> Today, the word law is used sparingly, if at all in most writings about evolution. Generalizations in modern biology tend to be statistical and probabilistic and often have numerous exceptions. Moreover, biological generalizations tend to apply to geographical or otherwise restricted domains. One can generalize from the study of birds, tropical forests, freshwater plankton, or the central nervous system but most of these generalizations have so limited an application that the use of the world *law*, in the sense of the laws of physics, is questionable. (Mayr 1982: 19)

Beatty (1995, 1997) has recently argued that distinctively biological generalizations, while true, cannot be laws because they are contingent on a particular historical pathway traversed as a result of evolutionary dynamics. Mendel's law of the 50:50 ratio of gamete segregation is true (when it is) only because the genes determining that ratio had been selected for in a particular episode in the evolutionary history of life on this planet. It could have been otherwise, hence is contingent on that particular evolutionary history and, for Beatty, is therefore not a law. Indeed, those historical conditions on which the truth of the generalization is contingent (e.g., those determining the selective advantage of the 50:50 segregation gene) may vanish in the future, rendering the generalization no longer capable of truly describing the state of nature. This feature of biological rules Beatty calls "weak contingency." In addition, by "strong contingency" Beatty denotes the situation in which from the same set of conditions with the same selection pressures operating variant functionally equivalent outcomes may be generated. Thus, one of the possibly multiple rules describing these variant outcomes appears not to be necessitated by the prior conditions that gave rise to it. "To say that biological generalizations are evolutionarily contingent is to say that they are not laws of nature – they do not express any natural necessity; they may be true, but nothing in nature necessitates their truth" (Beatty 1995: 52). Thus, the knowledge we have about life on Earth is victim to two failures of lawfulness. On the one hand, weak contingency violates the required universality in space and time. Strong contingency points to the exception-rich diversity of biological structures and processes that challenge the truth or necessity of any proposed evolutionary generalization.

However, the evolutionary contingency that Beatty attributes to biological generalizations does not separate out biological generalizations from those of the other sciences. All scientific laws or laws of nature are contingent in two senses. First, they are clearly logically contingent. Second, they are all "evolved" in that the relations described in the law depend on certain other conditions obtaining. That Galileo's law of free fall truly describes relations of bodies in our world requires that the mass of the earth be what it is. If, for example, the core of the earth were lead instead of iron, the quantitative acceleration would be four times what it is (though it would still be an inverse square relation). That the earth is configured the way it is is the result of the origin of the universe, the creation of the stars and planets. Generally stated, there are conditions in our world on which the truth of laws, such as Galileo's law of free fall, depend. They all could have been otherwise. This is the case whether or not those conditions are the result of particular episodes of biological evolution and are subject to further modification, or whether they are conditions that were fixed in the first three minutes of the birth of the universe. Whatever else one believes, scientific laws describe our world, not a logically necessary world.[3] All laws are logically contingent, and yet there is still a difference between Mendel's law of 50:50 segregation and Galileo's law of free fall. How can we represent that difference? Beatty is correct to note the difference, but incorrect to identify it with the contingency of the former. That there is a difference between Mendel's laws and Galileo's law should be explained, but it is not the difference between a claim that could not have been otherwise (a "law") and a contingent claim (a "nonlaw"). What is required to represent the difference between these two laws is a framework in which to locate different degrees of stability of the conditions on which the relation described is contingent. The conditions on which the different laws rest may vary with respect to stability in either time or space or both.

The dichotomous opposition between natural contingency and natural necessity in Beatty's discussion can be interpreted as a product of framing natural

---

[3] Sober allows that analytic statements, too, can be laws. Of course, there is an issue about what constitutes an analytic statement, that is, whether or not scientific commitments designate certain claims as methodologically analytic. Such matters require more attention that I am able to devote here. E. Sober, "Two Outbreaks of Lawlessness in Recent Philosophy of Biology," *Philosophy of Science* 64 (1997): S458–S467.

relations in logical terms. Logical necessity and contingency are indeed dichotomous alternatives. Yet imposing that feature onto the natural relations discovered by science limits what one can express about those relations. The difference between generalizations in physics and those in biology is inadequately captured by the dichotomy between necessity and contingency. They could both have been otherwise. What it would take to make them otherwise is different. They, therefore, have different degrees of stability. The condition that the material forming the core of the earth is iron, on which the strict representation of Galileo's law depends, is more stable in space and time than the conditions upon which Mendel's law rests. The actual acceleration of falling bodies, given those conditions, is deterministic, while Mendel's law is probabilistic. Thus, they differ both in stability and in strength.

With this framework in mind, what are the implications for the distinction between accidental truths and laws? That difference, too, will turn out to be a matter of degree and not kind. If we locate different true claims on a continuum, one can better see the distinctions. First, it becomes clear that even the so-called accidental generalizations are not all alike. (See Fig. 5.1.)

Think of the example of the coins in Goodman's pocket all being copper. We can formulate that in the standard model $(x)(Px \rightarrow Qx)$. For all things in the universe, if it is a coin in Goodman's pocket, then it is copper. Recall the counterfactual. If a coin were to be placed in Goodman's pocket, would it be copper? No. A quarter could easily be put in the pocket and falsify the universal generalization. What about the true universal claim that all spheres of gold occurring naturally on the earth have a diameter of less than 100 meters? What

---

Ideal Laws: Contingent, Universal, True

- Law of conservation of mass-energy
- Law of conservation of mass
- Second Law of Thermodynamics
- Periodic law
- No uranium-235 is larger than 55 kg
- Galileo's law of free fall
- No gold is larger than 55 kg
- Mendel's law of independent assortment
- All the coins in Goodman's pocket are made of copper.

Accidental Generalizations

Figure 5.1. Continuum of contingency.

conditions are responsible for this truth about our world? What would have to be different for the counterfactual to be false? That there is not sufficient gold in the desired configuration is something quite deep about the history of the universe and the distribution of matter. It is the result of the processes of stellar fusion and solar system formation. At the origin of the universe in the Big Bang sixteen billion years ago, it is believed that only helium and deuterium and lithium were formed within the first few minutes after that event. All the other elements, including the gold and uranium in our examples, have been produced in the subsequent evolution and development of the stars. (Beryllium and boron developed later as well, but by a different process.) The spectra of very old stars, which formed over 10 billion years ago, show deficiencies in all elements except hydrogen and helium, and so it is believed other elements have been synthesized since that time (Cox 1989). Thus, there is a sense in which it is impossible, given the history of the universe, that the so-called accidental generalization about gold would be false.

Considering these facts makes the gold example look more like the contrasting uranium so-called law (all spheres of uranium are less than 100 meters in diameter) than the so-called accidental truth about the coins in Goodman's pocket. What conditions would have to be different to undermine the alleged law about uranium spheres? If the ways in which the particles of uranium interact were changed, a sphere 100 meters in diameter would be possible. Note that this discussion is not about the conditions stated in the antecedent of the law statement. That is to say, it is not the Px in $(x)(Px \rightarrow Qx)$. Rather, I am referring to the conditions that underwrite the truth of the *relation* between P and Q described in the law. In the uranium case, for example, a different configuration, a sheet rather than a sphere, would permit a mass of uranium equal to that of a 100-meter-diameter sphere to occur. That would not constitute a violation of the law, but merely a non-Pa situation. On the other hand, the prohibited sphere of uranium does become stable at an extremely low temperature. Thus, if the history of the earth were such that the temperature never rose above that low level, then the so-called uranium law would not be true. Since the evolution of the Earth does include a higher temperature, uranium spheres with diameters of 100 meters are not stable.

The difference among the examples of Goodman's coins, gold spheres, and uranium spheres appears to be in the nature and degree of contingency they display: The conditions that allow the uranium law to truly describe our world are stable and connected with other causal structures in our world. The conditions that allow the gold law to be true are also stable, but less so: If we had "played the tape" of the origin and evolution of the universe again, it might have been the case that more of this element would have amalgamated naturally on the earth. Indeed, it may well be that such a configuration of gold may be found elsewhere in the universe.

Having displayed the variation of the nature of contingency of the standard philosophical examples, I want to now show how what scientists have identified as laws are also variable in the same way. At one end of the continuum are those regularities the conditions of which are stable over all time and space. At the other end are the so-called accidental generalizations. And in the vast middle is where most scientific generalizations are found. It is my view that to reserve the title of "law" for just one extreme end is to do disservice to science by collapsing all the interesting variations within science into one category: nonlaws. Indeed, by doing so we are unable to differentiate them from the least useful of the so-called accidental generalizations, for example, the coins in Goodman's pocket. By focusing discussion on laws versus accidental generalizations, natural necessity versus contingency, one is saddled with a dichotomous conceptual framework that fails to display important differences between the kinds of causal structure found in our world and differences in the corresponding scientific representations of those structures.

At the end closest to the philosophical ideal of exceptionless universality one might place conservation of mass-energy. Lavoisier published his law of conservation of matter in 1789. Leibniz had introduced a law of conservation of energy. In the early part of the twentieth century, with the acceptance of Einstein's theory of relativity the two laws were combined to express the view that mass and energy are alternative aspects of a single entity. Thus the law of the conservation of mass-energy is now understood to include both the mass of the matter in the system and the mass of the radiant energy in the system. In what sense is this law universal and exceptionless? Prior to the twentieth century the conservation of matter law would have been thought to be universal and exceptionless. Now we know that it fails to describe relations when energy levels are high enough to allow transformation of matter to energy. Yet do we now want to say that the conservation of matter is not a law? Scientists use it every day. It allows for reliable expectations in almost all circumstances. Conservation of mass-energy covers more domains. It is now believed to be applicable to all space-time where mass-energy is present. It is contingent on features of our world that are extremely stable. But even if there were regions where it failed to apply, say, near a black hole, it would be no less useful under a wide domain of application. Exceptionlessness is not required for a law to be useful as long as there is some understanding of its domain of applicability.

In a different location on the continuum one can place the second law of thermodynamics. What would the world be like for it to be no longer applicable? On the classical formulation it would be a world in which a perpetual motion machine could be built. One could extract work from a closed system of molecules without a corresponding increase in entropy. The statistical interpretation of the mechanism underlying this pattern describes the most probable distribution of molecules interacting with each other and the boundaries of the system. What would the world have to be like for the probable distribution to fail? Notice that even though rather basic features of our world ground this pattern, it is not exceptionless nor universal in the same degree as conservation of mass-energy. It is possible for entropy to spontaneously decrease. That means that all the matter in our world could be exactly the same, all the laws that apply be exactly the same, and still the second law would not accurately describe some part of the universe. It is possible that molecules left to their own devices would congregate in a structured way. But it is unlikely. In fact, one can calculate how unlikely it is. Thus, though it does not forbid an event occurring that contravenes the second law, the likelihood of it occurring is so small as to be negligible. There is "necessity" here in the sense of expectability, and thus the relations described by the law will obtain and hence one can reliably apply it. However, it is clearly not as strong as mass-energy conservation.

Continuing along, the exact formulation of Galileo's "law" of free fall (including the acceleration due to Earth's gravity) is conditional on the mass of the earth. If the earth was of a different mass, then it would no longer hold. And that the mass of the earth is what it is is a feature of the evolution of our universe: Had the core been lead instead of iron, then acceleration of free-falling bodies would be four times as great.

Now to Mendel's law. It states that with respect to each pair of genes of a sexual organism, 50 percent of the organism's gametes will carry one representative of that pair, and 50 percent will carry the other representative of the pair. The evolutionary history that makes such a description of our world true is, as Beatty has argued, a function of a particular set of complex circumstances that allowed a genetic mutation for 50:50 segregation to appear in an environment where it was more fit than other variants, in a context where other features of the organisms were not traded off via other selective pressures, where chance and mutation and migration did not override the selective advantage, and so on. If any one of those sets of conditions had been otherwise, a different genetic evolution might have occurred and the law would find no application in predicting, explaining, or intervening in our world. That is, given that the circumstances did occur, Mendel's law operates to perform the functions we require of laws. Putting these examples on a continuum of stability, we get the following results:

Is conservation of matter not a law if it fails to apply in nuclear reactions? No. That most of the universe's physical systems are structured in a way that accords with the relation described lets us use the "law" to explain and predict successfully. Does Mendel's law of 50:50 gamete segregation fail to be a law because it doesn't apply to the entire temporal period prior to the evolution of sexual reproduction nor to cases where a mutation for meiotic drive changes the relative fitness ascriptions? Where the requisite conditions hold, Mendel's law also allows us to explain and predict successfully.

The difference, then, between the two is not that one functions as a law and the other does not, nor that one is necessary and the other is contingent. Rather, the difference is in the stability of the conditions on which the relations are contingent. Consequently, there is a difference in the kind of information required in order to *use* the different claims. It would be great if we could always detach the relation discovered from its evidential context and be assured it will apply to all regions of space-time. But virtually no scientific law is so unrestricted. Whether it is in space or in time, there are regions where the relations described will hold and regions where it will not hold. The conditions that describe those regions are the conditions upon which the truth of the relations is contingent. To apply less than ideally universal laws, one must carry the evidence from the discovery and confirmation contexts along to the new situations. As the conditions required become less stable, more information is required for application. Thus, the differences among the laws of physics, the laws of biology, and the so-called accidental generalizations are better rendered as degrees of stability of conditions on which the relations described depend, and the practical upshot is a corresponding difference in the way in which evidence for their acceptance must be treated in their further application.

How can one represent these degrees of stability and strength? Philosophers have certainly developed alternatives to the first-order predicate calculus to represent regularities (Skyrms 1980; Spirtes et al. 1993; Suppes 1974; Woodward 1997). The probability calculus is the most obvious. More recent developments include causal graphs and computer models. These alternative representations don't suffer some of the limitations of standard formal logical systems. For some of the causal structures that scientists might refer to as "laws," these are better representations. For example, they are better representations of "laws" of thermodynamics and of economics. For some causal structures, the additional representational machinery appears not to be necessary, for example, the law of mass-energy conservation. What's the lesson? The lesson is that one representation really can't capture the structural features of all the kinds of causal structures found in nature. But this result is only the first step in accepting that a characterization of "laws" is going to have to be complex.

The next step is the recognition that no structural representation of the degree of abstraction provided by the first-order predicate calculus, the probability calculus, or causal graphs is sufficient to completely characterize a causal structure for the purposes of explanation and prediction. Why? Because they do not include the conditions of applicability of the laws when they are used for these purposes. Consider the ideal gas law: $PV = nRT$. As every elementary text says, this law applies when "the energy of interaction between the molecules is almost negligible compared to their kinetic energy" (Reif 1987: 176). In an actual laboratory measurement setting, even that condition of applicability has to be translated into a particular situation in terms of the kind of gas, the temperature, and the pressure in question. There is also the obvious fact that the law does not apply as the number of molecules of the gas in question becomes small. In short, there are a very large number of characteristics of this "law" qua claim about the world that are not captured in an algebraic equation (or the first-order predicate universal form of that equation). These characteristics are relevant to the applicability of this claim for explanation and prediction. These characteristics, I claim, are constituents of a complex characteristic space that is the appropriate analytical framework from which to understand any particular scientific claim and from which to understand the relations between the various kinds of claims that scientists make. This complex space includes a number of characteristics that determine the applicability of a law. Many of these characteristics are multi-valued or even continuous-valued. Thus it has both more degrees of freedom and more complex descriptive parameters than the traditional accounts.

Initial approaches to describing some of the dimensions of the characteristic space for generalizations have already been taken. Skyrms and Woodward have recognized differences in degree of lawfulness in the claims of science. Both describe the lawful character of generalizations by appeal to the resiliency of the relations described by the claim. Within their intended domains, described relations will exhibit some degree of resiliency: The probability of the consequent condition on assumption of the antecedent condition will be more or less stable. For universal generalizations, the probability is one that within the specified domain no counterexample will be found. If it is maximally resilient, then it will be exceptionless. But it may be less resilient than that and still afford us reliable expectations of what will probably occur. Hence, Skyrms allows for the representation of less than ideal lawful relations. The variation he permits is in the strength of the relation (and, hence, corresponding expectation) between the factors in a specified domain. Woodward develops his concept of resiliency as a condition for explanatory power by appeal to the kinds of interventions one could make in a setup and still have the generalization correctly describe the results. The more resilient, the fewer interventions will issue in a failure of the relation to hold. Thus, Woodward explores the domains of applicability that permit the lawful relation to hold. With these refinements, Skyrms and Woodward want to retain a law/nonlaw dichotomy. Skyrms says:

> ... the more nomic force, the more central the law is to our conceptual scheme, that we are less willing to give up a law than an accidental generalization, that giving up a law is more disruptive to our conceptual scheme than giving up an accidental generalization, and the more nomic force the greater the disruption. ... so I would say that it [no gold sphere has a diameter more than 100 m] has more nomic force than the generalization about coins in Goodman's pocket on V. E. day, but less than anything we would regard as a genuine law.

My view differs in at least positing that what makes it more or less difficult to "give up" a claim in any given circumstance is the nature of the conditions under which the generalization is applicable. Skyrms's and Woodward's analyses describe some dimensions of the multidimensional space of characteristics that more adequately represent laws and their applicability. To fill out the rest of the details is to give a pragmatic view of laws and their operation in science.

The Pragmatic Strategy

Taking a pragmatic approach to scientific laws replaces a definitional norm with an account of the *use* of scientific laws. How do they function to allow us to make predictions, explanations, and successful interventions? The result is a framework for representing the multiple types of generalization found in the sciences. In contrast to the dichotomous space defined by the normative approach, this view requires a multidimensional frame in which to view the varied conditions of applicability of scientific generalizations.

Scientists search for knowledge of the causal structures in our world. When we know what sorts of properties or events are causally relevant to the production of other properties or events, then we can use that knowledge in pursuing both scientific and practical ends. Again, it would be ideal if we could always detach the generalizations gleaned from specific investigations from their supporting evidence, carry these laws to all regions of space-time, and be ensured of their applicability. Such generalizations would be universal and exceptionless. But some causal structures – in particular, those studied by biology and the social sciences – are neither global nor exceptionless. Thus, the generalizations describing them cannot be completely detached from their supporting evidence. Nevertheless, we can and do use these more limited tools to do the jobs we set out to do. To know when to rely on a generalization that does not apply to all space and time we need to know when it will apply, and this can be decided only from knowing under what specific conditions it has applied before and the caveats its mode and manner of representation warrant for explanatory and predictive applications.

In my 1997 paper I distinguished between two general types of parameters that structure the applications of scientific regularity claims: ontological and representational. Scientific knowledge consists of claims about the causal structure of the world and at the same time is represented in some form, be it linguistic, mathematical, or visual. The complexity that is reflected in the diversity and plurality of claims in the sciences reflects ontological differences both among the causal structures in the domains studied and in other features of the representational medium.

The ontological parameters include what I have been calling the continuum of stability of the conditions on which the causal relation depends. As we saw, in the case of Mendel's law of 50:50 segregation, that relation will hold just so long as the genetic and environmental conditions persist that render 50:50 segregation adaptive. Much can happen to perturb those conditions, including the introduction of a gene that induces meiotic drive (while at the same time not being coupled with other maladaptive effects, such as sterility as in the case of the T-allele in the house mouse). With the second law of thermodynamics, one ontological requirement is that our subject be a system with a large number of molecules.

The second ontological parameter is what I have called a continuum of strength and it refers to the relation described by the arrow in $(x)(Px \rightarrow Qx)$. It refers to the difference in strength between a deterministic law and a probabilistic one. It seems clear that one of the ways in which biological laws have been understood to fail to meet the ideal standard is by being nondeterministic. Mayr clearly means this when he says, "Generalizatons in biology are almost invariably of a probabilistic nature. . . . [There is] only one universal law in biology 'All biological laws have exceptions'" (Mayr 1982: 38). In addition to the ontological parameters, I suggest above that other parameters are also relevant to the applicability of generalizations. These are degree of accuracy, level of ontology, simplicity, and cognitive manageability. In addition we need to consider the degree of abstraction most appropriate to the pragmatic goals of inquiry. This latter consideration acknowledges that some patterns of causal interaction will be visible only when certain details are ignored.

I do not claim this to be an exhaustive list. In fact, I believe better characterizations of "laws" will require both filling out the details about the characteristics listed here as well as extending this list. I have already described in some detail the characteristic of stability and the characteristic of strength. I now turn to a discussion of the next characteristic on the list: degree of abstraction.

With respect to degree of abstraction (see Cartwright 1989), one sees that different levels are required for different tasks. For example, Darwin's insight into the causal patterns responsible for adaptedness was a reflection on the similarity between the varieties of cowslip and English game pigeons. Nearly all material properties of the two populations are different, except that they are made of organic molecules and have DNA and mass. What was seen as similar was the fitness of the surviving members of the populations to their respective environments. To explain the adaptedness of species, a feature that had seemed indisputably the result of the design of the creator, Darwin detected a pattern in the various populations that had a single, mechanical explanation. Natural selection operating on variations would cause those with any slight advantage to persist, whether the advantage be taller stalk, thicker beak, or darker pigmentation (see Mitchell 1993). The schematic character of this "law" has been emphasized by Brandon (1978) and others, recognizing the great deal of "local information" required to apply the knowledge of evolutionary causal structures to concrete cases.

In chemistry, there are a variety of configurations of molecules that we identify as water or as hemoglobin.[4] Representing these differences is useful for different purposes. We can treat water abstractly to refer to all isotopes of H_2O as well as configurations that replace the hydrogen with deuterium, which has one neutron, and tritium, which has two (hydrogen has no neutrons). They all have one proton and one electron. So we could have T_2O, DOT, HOT, and so on. There are two isotopes of oxygen, O-17 and O-18. In all, there are twelve possibilities. When can we abstract away from these variations and treat a situation as one containing water and describe usable causal regularities? Replacing hydrogen in a molecule with deuterium can slow down the rate of the reaction. Deuterium is almost twice as heavy as hydrogen, so it moves slower. Deuterium has different spectroscopic properties than hydrogen; thus, their resonance properties vary. This difference is useful to represent in the context of proton nuclear magnetic resonance contexts. For example, such procedures are often performed in deuterated solvents, so that

---

[4] I owe these examples to Michael Weisberg.

the only hydrogen in a sample will be from the molecule being analyzed. Thus, replacing $CHCl_3$ with $CDCl_3$ removes hydrogen from the surrounding medium to allow detection of hydrogen only in the test sample. But for studying tidal properties of the ocean, for example, representing these differences would be confusing and unnecessary.

The case for hemoglobin is even more complex. Hemoglobin contains 2,954 carbons, 4,516 hydrogens, 780 nitrogens, 806 oxygens, 12 sulfurs, and four irons. There are three natural isotopes of carbon, two of nitrogen, two of oxygen, four of sulfur, and four of iron. The number of possibilities for different isotopes is so large that there are almost certainly no two identical hemoglobin molecules in an individual's body. This is the case even when we consider that one drop of blood contains 10^{17} molecules.

If we make the simplifying assumption that all isotopes are equally probable, the number of different hemoglobin molecules is a number with 4,132 decimal places. That is, the order of magnitude is $1 \times 10^{4,132}$. This means there are about 5×10^{20} different kinds of molecules in an individual's body (assuming no two are the same). To collect all the possible hemoglobin molecules, we would need at least $2 \times 10^{4,111}$ people. Of course, there are only about 5×10^{9} people in the world.

When would we want to represent the molecules in sufficient detail to capture these differences? When is the level of abstraction that refers to all isotopes as hemoglobin *simpliciter* adequate? These questions can be answered only with reference to a particular context and scientific objective. For most chemical reactions these differences are negligible, since the chemical behavior of a molecule is a function of its shape and electronic configuration. The variations between isotopes in electronic structure are negligible. The level of abstraction we need to represent a situation is determinable only by the problem we wish to solve – the use to which the knowledge is to be put.

Different tools will be better at solving different problems. What we should expect to see in the sciences is a diversity of models, theories, and levels of abstraction. And indeed, Feynman describes the heuristic benefit of multiple representations of empirically equivalent theories:

> ... psychologically we must keep all the theories in our heads, and every theoretical physicist who is any good knows six or seven different theoretical representations for exactly the same physics. He knows that they are all equivalent, and that nobody is ever going to be able to decide which one is right at that level, but he keeps them all in his head, hoping that they will give him different ideas for guessing. (1995: 168)

That different representations may be better suited to different tasks is obvious. If I want to navigate in the city of Washington, D.C., to find the Capitol, it will be equally unhelpful to have either a one-to-one full-scale representation of the city and its buildings or a map of the United States where Washington is represented as a single point.

The details of the rest of the list of characteristics in the complex space of the pragmatic account of laws will have to be left for future work. But having the general account and some specifics about the characteristics of stability, strength, and degree of abstraction addresses the original conundrum: How can there be no "laws" in biology when biologists think they use laws all the time to explain and predict?

The failure of knowledge claims in biology or other sciences to live up to the universal, exceptionless character of the ideal case does not preclude their functioning as "laws," generalizations that ground and inform expectations in a variety of contexts. In the multidimensional space defined by the multiple aims of scientific practice, including the ontological and representational parameters, it may well turn out that all or most of the generalizations of physics occupy a region distinct from the region occupied by generalizations of biology (see Fig. 5.2). The conditions on which physical laws are contingent may be more stable through space and time than the contingent relations described in biological laws. The conservation of mass-energy law is

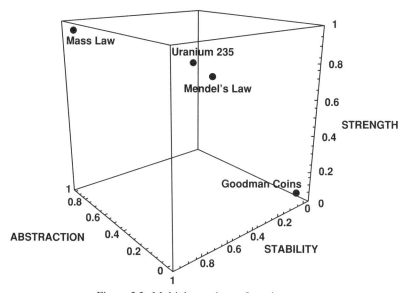

Figure 5.2. Multiple continua of contingency.

more stable than Mendel's law of segregation. The strength of the determination can also vary from low probability relations to full-fledged determinism, from unique to multiple outcomes. Indeed, the causal contribution of particular features may vary depending on the features' sensitivity to environmental conditions, their resilience in Skyrms's and Woodward's senses. In terms of abstraction, Mendel's law may be considered more abstract than the law governing the possible configurations of uranium. While I have only sketched the parameters by which knowledge claims may be compared, it is clear that such a conceptual framework has the resources to display the multiple relationships that exist among and between generalizations in the sciences. The world is complex and so must be our scientific representations of it. So, too, for the world of scientific knowledge.

5.3. CONTINGENT GENERALIZATIONS: LESSONS FROM BIOLOGY

Introduction

Biological systems are characteristically complex.[5] They are historically contingent, having arisen in a path-dependent history of dynamic change over both random and nonrandom components. Biological systems are generally made up of many parts that stand in structured relationships. Take an individual organism, such as a human being with its approximately 210 different cell types and its more than 30,000 genes, or a honeybee colony with its tens of thousands of individual workers engaged in different, coordinated tasks. Such systems are describable in terms of multiple levels of organization from gene to cell to organ to organism to social group. And these levels interact in a variety of ways with each other. In addition, such systems tend to be subject to multiple, interacting causes rather than a single, dominant cause at each of those levels of organization. In short, they are multicomponent, multilevel interacting systems. It is my view that the complexity of the systems studied by biology and other sciences has implications for the scientific knowledge we should aim to generate in order to represent such systems. In this section I explore the implications of complexity for contingency and suggest some methodological consequences of different types of contingency.

Contingency has a special significance in understanding biological systems. Recently, Beatty (1995, 1997) identified *evolutionary* contingency as the characteristic of biological generalizations that make them fail to conform

---

[5] This section was originally published as S. D. Mitchell, "Contingent Generalizations: Lessons from Biology," in R. Mayntz, ed., *Akteure, Mechanismen, Modelle: Zur Theoriefähigkeit makrosozialer Analysen* (Frankfurt: Campus, 2002), 179–195.

to the criteria for natural law: "To say that biological generalizations are evolutionarily contingent is to say that they are not laws of nature – they do not express any natural necessity; they may be true, but nothing in nature necessitates their truth" (Beatty 1995: 52).

On this view physics is a lawful science but biology is not. This is a significant deficit for biology, since laws are thought to play a special role in science. They are what science allegedly seeks to discover. They are the codifications of knowledge about the world that are supposed to enable us to explain why what happens happens, to predict what will happen in the future or in other circumstances, and, therefore, to intervene in the world in order to reach our pragmatic goals.

I have argued elsewhere (Mitchell 1997, 2000) that claiming that biological generalizations are contingent is insufficient to either distinguish them from truths of physics, which also share some forms of contingency, or clearly mark what is unusual about knowledge claims in biology. In these earlier papers I suggested that contingency comes in degrees so that the difference between generalizations in biology and in physics is not one of a lawless and lawful science, but rather a difference in the degree the causal dependencies described depend on the prior conditions. I further suggested that scientific laws have many dimensions by which we can judge their usefulness. Two main features of general scientific knowledge claims are: (1) ontological content, that is, the nature of the systems that they represent, and (2) pragmatic attributes: how usable this knowledge is for addressing the problems we want to solve, for matching our computational abilities, and so on. The complexity of biological systems affects both these dimensions.

The intuitive notion of the contingency characteristic of evolutionary biology is expressed in Steven J. Gould's metaphoric appeal to Frank Capra's *It's a Wonderful Life* (Gould 1989). That is, if we rewound the history of life and "played the tape" again, that which would turn out as species and body plans and phenotypes would be wholly different. The causal structures that occupy the domain of biology are historical accidents, not necessary truths. Not so for the laws of physics, such as Newton's gravitational force law. That law describes the force acting on two bodies as a function of the gravitational constant, the masses of the two bodies, and the distance between them. Such dependence is taken to be universal, exceptionless, and physically necessary. As long as there are two bodies in the universe, the gravitational force law will apply. As long as there are sexually reproducing species on the planet, will Mendel's law of segregation apply?

According to Mendelian genetics, diploid organisms, like human beings, have two alleles or genes at every locus on a chromosome that determine the character of a phenotypic trait, such as eye color. Mendel's law of segregation states that during gamete formation, each member of the allelic pair separates from the other member to form the genetic constitution of the egg or sperm. This is done by the process of meiosis. Mendel's law of segregation is generally true, but exceptions, though rare, occur when meiotic drive takes place. Meiotic drive can be accomplished either by preferential segregation during meiosis or by lowered recovery (through death or dysfunction) of meiotic products that do not carry the driven gene (reviewed by Zimmering et al. 1970). In such cases a gene will double or triple its gamete representation, making it more likely than the other half of the pair to be represented in the offspring. Thus, the operation of Mendel's law is contingent not only on the existence of sexually reproducing organisms, but also on the absence of conditions that produce meiotic drive. There seems to be a sense in which Mendel's law is more contingent than Newton's. One of the aims of this paper is to clarify what is meant by this sense of contingency.

Causal dependencies that hold in our world do not do so because of language or logic. They do so because that is the way our world works. This is as true for thermodynamics as it is for population biology. Nevertheless, many of the relationships connecting physical properties are more invariant than are the relationships connecting biological properties. There is a difference between fundamental physics and the special sciences, but it is not the difference of a domain of laws versus a domain of no laws. Only by broadening the conceptual space in which we can locate the truths discovered in various scientific pursuits can we come to understand the nature of those differences. It is true that Mendel's law is contingent, not necessary. But contingency comes in many degrees and types. One type of contingency, logical contingency, is a feature of all causal dependencies, in fundamental physics and biology alike. So just saying that a dependency is contingent doesn't tell you much at all, and certainly doesn't characterize what is distinctive about the causal structures in complex systems studied by the special sciences.

What I do in the remainder of this section is explore different types of contingency in social insect biology. It will be apparent that the dichotomy between law and nonlaw, or necessary and contingent, which has been used to mark the difference between the knowledge of fundamental physics and that of the special sciences as illustrated by the example of biology, is inadequate. In the end, I draw some implications for the methods appropriate to study the contingent regularities that characterize complex systems.

My approach looks to some new philosophical developments in our understanding and representation of causality. These developments focus on how to make use of less than ideal knowledge, for example, how to draw causal inferences from statistical data, as well as how to determine the extent of our knowledge from looking to the practices of experimentation (Spirtes et al. 1993; Woodward 2002a). Thus, rather than descending from the lofty heights of God's laws or LaPlace's demon's-eye-view of the way nature is and deciding what our knowledge should look like on the basis of such presumptions, we start where we are. When we have only statistical correlations, what can we know? When we manipulate experimental setups, what do we know about natural setups? When we see that some relationship holds here now for this population in this environment, what can we say about the next population in the next environment?

What is interesting about the new approaches is their ability to represent the less-than-universal character of our knowledge of many causal structures we find in the world. Of course, there are two interpretations possible. The first is a realist, ontological thesis that causal structures in our world are *not* necessarily universal and exceptionless, and so representations of them as such are not really an appropriate end product of scientific investigation. The other is an epistemological view that although there are universal, exceptionless causal laws underlying the causal dependencies we detect, we have to make do with our knowledge of the current, less-than-universal, exception-ridden generalizations until we discover the universal laws. My own view is that there is not much difference in the day-to-day practice of science and the methodological strategies one adopts for the two views. On either interpretation, we can agree that it would be much easier if when we detect a system in which A was correlated with B, we would know not just that A caused B in that system but that A would cause B in every system. When we look at the tidy behavior of Mendel's pea plants, where internal genetic factors assort independently, we would be better off if we could say that that would always happen in any sexually reproducing population. But it doesn't. And since it doesn't, we need to understand more about systems of Mendel's peas and their relationship to other systems to know which aspects of the original test case are exportable to the new domains. If we were so lucky as to have the ideal universal exceptionless relationships of the philosopher's strict law, we would know it would automatically apply to all times and all places. But that is not the world of our scientific practice.

Especially in the scientific domains that display multiple contingencies, where the behaviors of the systems depend on specific, local configurations of events and properties that may not obtain elsewhere, and that rely on the interaction of multiple, weak causes rather than the domination of a single, determining force, the generalizations we can garner will have to have much more information accompanying them if we are to use that knowledge in new contexts. The central problems of laws in the special sciences is shifted from "Are they necessary or contingent?" to "How do we detect and describe the causal structures of complex, highly contingent, interactive systems, and how do we export that knowledge to other, similar systems?" This is where complexity and contingency join forces to shape the types of scientific generalizations that are appropriate to the special sciences.

Types of Contingency

Logical Contingency. The causal structure of our world is not logically necessary. Causal relations in physics as well as in biology depend on some features peculiar to our world. As I have argued (Mitchell 2000), all truths about our world that are not logical or mathematical truths are contingent on the way our universe arose and evolved. This is as true for laws of physics, which describe causal structures that might have been fixed in the first three minutes after the Big Bang; laws of atomic structure that came to apply as the elements themselves were created in the evolution of stars; and the dependencies among genes and phenotypes, the rules of which changed as life on the planet first formed out of the primordial soup, replicated in a single-celled existence, and became more complex and differentiated as multicellularity, sexual reproduction, and social groups subsequently evolved. All causal dependencies are contingent on some set of conditions that occur in this world but not in all possible worlds.

Space-Time Contingency. The conditions on which causal structures depend are not equally well distributed in space and time (see Waters 1998). Mendel's second law of independent assortment of gametes was empty until the evolution of sexually reproducing organisms. It is believed that sexual reproduction evolved as early as 2.5 to 3.5 billion years ago (Bernstein et al. 1981). The earth is 4.54 billion years old, so it was about one billion to two billion years before sexually reproducing organisms arose and the "rules" governing their behavior, the causal dependencies or causal structures that they carry, were operant. But universality never meant that the causal structure described in a law would occur at every point in space-time. Rather, whenever and wherever the conditions for a relationship between the properties did occur, it would occur by the relation described by the law.

Weak Evolutionary Contingency. Evolved and evolving systems display a form of contingency that the structures that were fixed early and forever in the development of the universe do not. Such evolutionary contingency is described by Beatty (1995) as "weak evolutionary contingency." By this, he

151

means that the conditions on which a particular causal structure depends (sexually reproducing organisms and absence of conditions for meiotic drive for Mendel's law, for example) may change over time. It is the case that biological structures come and go. While some are conserved from the beginnings of life until now, others appear and then are lost, while others appear only later in time. Any evolved structure conceivably could disappear if the set of conditions that maintains it as evolutionarily adaptive and developmentally stable changes. This is a case not just of frequency, as in space-time contingency, but of duration. Once uranium, for example, was synthesized in the development of stars, at least six million years after the Big Bang, its molecular structure was set for all time (Cox 1989). But in biology, features, species, and even entire taxa have been lost over time as organisms evolve. Body plans, mating strategies, and other organized structures flourishing at one time are completely absent at another time. Even the very features we take to be diagnostic of the character of an order – for example, the backbones of the vertebrates – could in the future be selected out, and the heirs of vertebrates may have no backbones at all! Thus, the causal dependencies describing vertebrate ontogeny may well be restricted in temporal domain. And although the causal dependencies describing uranium molecular structure also are restricted to a temporal domain, that is, six million years after the Big Bang, the degree of restriction to the past and future is clearly less substantial.

Strong Evolutionary Contingency. Beatty (1985) also identifies "strong evolutionary contingency," which is a feature of complex or chaotic systems. Contingency in this sense occurs when under the "same" starting conditions, different resultant effects occur. Thus, according to Beatty, given the same set of conditions with the same selection pressures operating on a population, variant functionally equivalent outcomes may be generated. For deterministic chaotic systems, this is also the case. Insofar as the initial conditions cannot be distinguished, no matter how precise our measurements become, radically divergent future states can be generated. The diversity of life on the planet attests to the prevalence of strong evolutionary contingency: Different "solutions" to evolutionary "problems" suggest there is no strongly determined outcome for a given evolutionary scenario. Thus, the particular solution, of the multiple ways possible, that a particular population adopts is contingent on the operation of chance factors, sensitivity to small variations in initial conditions, and dynamic interactions. When two perfectly inelastic bodies collide, a unique solution is determined. Not so for complex evolved biological systems.

Multilevel Contingency. In addition to the evolutionary, historical contingencies that are discussed for biological systems, there are other types of contingency that arise from the complex organizational structure of these systems. These are the result of the ways in which both the multilevel structures and the multiple components at a single level may interact to produce phenomena. There can be suppression of the "normal" operations of causal mechanisms at a lower level by the control exerted by its being embedded in a hierarchical organization. For example, in a honeybee colony, the normal ovarian development in females is contingent on the presence or absence of a queen. When the queen is present, a pheromone is emitted that stops almost all female workers from developing ovaries. When a queen dies or is removed, the development of ovaries resumes. Thus, there are organizational contexts that condition the operation of particular causal mechanisms. Much of the developmental system of honeybees and solitary bees may well be the same, but the operation of this conserved mechanism is contingent on the location of the system within a more complex structure.

Multicomponent Contingency. Another characteristic of complexity is the interaction of multiple causal mechanisms in generating a behavior of the system. The multiplicity of factors, though perhaps computationally challenging, is not particularly problematic, especially if there are simple rules of interaction, such as additivity in the case of gravitational and electromagnetic components. However, complex systems often involve feedback mechanisms resulting in the amplification or damping of the results of interacting mechanisms.

An example of this is found in the generation of foraging behavior in honeybees. It is understood that genetic differences among individuals account for some of the difference in foraging frequency. Although the specific pathways are unknown, the correlation between genes and foraging frequency is robust and, in the absence of other interacting factors, explains differences within and between colonies. However, it is also known that individuals change their foraging frequency as a result of environmental stimuli. In a genetically homogeneous population, variation in foraging frequency would then depend on the different environmental factors individuals had encountered. In such ideal situations where only one factor is active, a systematic study can reveal the strength of genes or the strength of environmental stimulus in producing the effects. However, in natural settings both genetics and learning are simultaneously operant. The result of their joint operation is not a simple linear compound of their individual contributions. Indeed, research has indicated that the results are amplified by the interaction of these two mechanisms.

What type of contingency is expressed here? In the multilevel contingency discussed above, the presence or absence of the mechanism was conditioned on its being located in an embedding system. In the case of multicomponent contingency, the very function describing the causal mechanism is altered if it is in a feedback system.

Rather than describing how the operation of component mechanisms is contingent on their interactions, it might be tempting to say that there is a new, wholly different mechanism in operation in the case of feedback, one that is described by the complex, nonlinear function. If one takes that path, then there would be three different mechanisms: genetic determination of behavior, environmental stimulus determination of behavior, and the feedback system of genes plus environment. However, this way of analyzing the situation ignores the very methods of decomposition that are used to uncover the causal dependencies that produce phenomena in complex systems. Indeed, both the evolution and development of complex biological systems support a strategy of decomposition. That is, social insects evolved from solitary ancestors. Much of the internal biology of the individual bee is preserved in this process. By understanding complex feedback structures as built from the interaction of component mechanisms, this fact can be respected. The mechanisms found in solitary ancestors are the same as those found in their social descendents, but their operation is contingent on the new context of social life. Thus, with the evolution of sociality, new contingencies emerge, not new primary mechanisms.

So, too, with developmental systems. Complex organisms as well as organizations develop from simpler preceding states. Diversification and stratification of cell type in the organism, or behavioral task group in the social insect colony, can issue in both upstream and downstream processes, turning on and off gene expressions or individual biochemical mechanisms. Yet it would be peculiar to suggest that once fully developed, the complex system operates by means of causal processes wholly different from the component processes that characterized its earlier stages. It is more in tune with our understanding of evolution and development to analyze these situations in terms of the contingent integration of component mechanisms.

Redundancy and Phase Change. Redundancy and phase change phenomena can also be present in complex biological systems. These offer two more types of complex contingencies that have import for understanding the range of contexts in which the regular behavior of a set of variables may be disrupted. In systems with redundant processes, the contribution of any one may be elicited or moderated by the operation of another. For example, redundancy of systems to generate a functional state will make experimentation in the standard sense less definitive. What could we learn from a controlled experiment? Imagine we are trying to determine the consequences of a particular component, say, of visual information on the direction of flight of honeybee foragers. We create two study populations that are identical for genetics, age, environment, food source, and so on. We block the visual uptake of the individuals in the first population, while leaving the system operative in the second population. Then we look to see what differences there are in the foraging behavior, ready to attribute differences to the role of visual information. It could well be that the foraging behavior in the two populations is identical. Does that tell us that visual information plays no causal role in foraging decisions? No. It is plausible to postulate that when visual information is not available, an olfactory system takes over and chemical cues in that mechanism generate the same behavioral responses – that is, ones that are adaptive or optimal responses to foraging problems.

Redundancy of mechanism makes controlled experimental approaches problematic. This is not to say that each modality or mechanism cannot be studied to determine the contribution of each in the absence of the activation of the others. However, redundancy can take a number of different forms that make the ways in which the mechanisms mutually contribute to an overall outcome vary. For example, as described in the hypothetical case above, there may be serial and independent redundant "backup" systems. There may be mutually enhancing or amplifying systems or, conversely, mutually dampening systems. Knowing the causal laws that govern each process in isolation by experimentation will not automatically yield sufficient information for drawing an inference to their integrated contribution in situ. As I discuss below, this type of complexity introduces a special set of contingencies that could apply to the operation of any partially isolatable mechanisms.

In phase changes, like those in nonlinear dynamical processes, the nature of the function describing the behavior of variables may itself change under certain values of the variables or changes in external conditions. This also affects epistemological practices. Consider situations whereby the causal relationship governing the behavior of two variables changes completely at certain values of one of the variables. Consider the slime mold, *Dichtyostelium*, which lives much of its life as a collection of individual single cells moving through space in search of food (Kessen 2001). The movements of the cells are driven by the detection of food. However, should a group of cells find itself in a situation in which the value of the food variable is below a threshold, that is, what would be close to starvation rations, then an entirely new set of causal dependencies kick in. Now, instead of each cell moving toward food in a predictable way, the individual cells are drawn to each other, amass together and form a new organism, a multicellular slime mold made up of a stalk and fruiting body. The latter emits spores to search for more nutrient-friendly hunting grounds. The rules governing cellular behavior have changed.

Which rules apply and when the rules apply is contingent. Complexity affords different kinds of contingency that must be understood to accommodate both the knowledge we have of complex systems and the practices scientists engage in to acquire this knowledge. I have analytically differentiated a variety of types of contingency that can affect complex biological systems. It is the case, however, that in the natural world these types of contingencies can occur simultaneously. This obviously poses methodological challenges for the researcher who wants to discover true generalizations that explain, predict, and permit us to intervene successfully when complex systems are involved. By way of a detailed example, I illustrate these features and then draw some methodological conclusions about how the complexity and contingency of biological systems influence both the conduct and results of scientific investigation.

Multilevel, Multicomponent Systems: The Case of Division of Labor

A colony of honeybees can contain a single queen and between 10,000 and 40,000 female workers (Winston 1987). These individuals engage in a variety of tasks over their lifetimes of four to seven weeks, including cell cleaning and capping, brood and queen tending, comb building, food handling, guarding, and foraging (Seeley 1982). This system provides a paradigm case of a multilevel, multicomponent, evolved complex system. The individuals are not randomly engaged in tasks, nor does each individual do all the jobs available. Rather, there is a pattern of distribution of behaviors that changes through the life of the bee. One aspect of the pattern is called "age polyethism." As the insects age, the type of work they do changes. Four age "castes" have been distinguished: cell cleaning, brood and queen care, food storage, and foraging. Each "caste" consists of a repertoire of tasks, and the individuals vary in their degree of specialization within a caste set. In addition, the colony as a whole displays plasticity by adjusting the number of workers engaged in the various tasks in response to both internal and external factors.

Let's suppose we want to explain or predict the state of the system at a particular time. Which generalizations govern the production of division of labor such that under certain initial conditions, the colony will have a colony-level distribution of 50 percent of the individual foragers resting and 50 percent foraging? Which causal factors play a role in determining this colony-level phenomenon? It is clear that there are multiple levels of descriptions that can

L1: Colony level [distribution of behaviors]

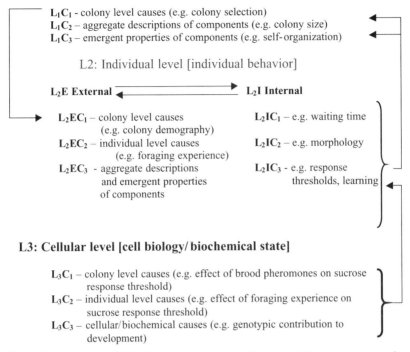

Figure 5.3. Division of labor in social insects: multilevel, multicomponent, complex system.

be invoked in such a case, and that at each level there are multiple causes that could be operant and interacting (see Fig. 5.3).

Starting at the colony level, one can characterize the phenomenon as a distribution or frequency of foraging, that is, 50 percent foraging, 50 percent resting. At that level of organization, the causal factors that operate *on* the colony include past colony-level selection for the ergonomic efficiency of this type of organizational structure (Wilson 1971). It has been argued that having some division of labor was adaptive relative to colonies with no division of labor in past environmental circumstances and that this explains its presence in the various ant, bee, wasp, and termite species where it does occur. Indeed, Wilson (1971) suggests this level of sociality in insects evolved independently eleven times. The applicability of this generalization is contingent on there having been a selection history that included the appropriate variation in colony-level traits and heritable components such that through differential replication (colonization and colony death), those with division of labor were more successful, and hence the trait persisted in subsequent generations.

Another cause of the same colony-level phenomenon is gross size (Pereira and Gordon 2001). Some studies have shown that, under certain assumptions of what cues workers use to govern task choice, the larger the size of the colony, the quicker the response time. So if the phenomenon to be explained is a rate of change to 50:50 foragers, then colony size can be a causal factor. It also seems to be the case that the larger the colony size, the better a colony can fine-tune its distribution of workers to internal and external needs. Certainly, it is a necessary, though not sufficient, condition that the more components in a system, the more differentiation is possible, and hence the more fine-tuning that is possible. If there is only one individual (or one cell), only two states are possible (100% foraging or 100% resting). With two individuals, only three states are possible (100% foraging, 100% resting, or 50% engaged in each task). With more individuals, more differentiated states are possible.

Other causal factors that contribute to the colony trait, besides selection at the colony level or size, are the result of compositional functions of the individual parts of the colony. That is, the 50:50 distribution of workers could be the result of mere aggregation of the individual behaviors of the workers. In that case, the causal factors determining it would occur at the next level down. Or it might be that the distribution is an emergent property from the interactions of the individual workers. Again we need to drop down to the individual worker bee level, but the means of moving back up to the colony level would be different. There are different composition functions describing how the parts constitute or cause the trait of the system. Total number of nest mates clearly will be an aggregate property of the system. Distribution of workers into different tasks has also been explained as the result of self-organization of the individuals (for a review, see Beshers and Fewell 2001; see also Mitchell 2002c and Page and Mitchell 1998). In the latter case the distribution is the nonadditive result of individual interactions with each other and the external environment. Thus, whether a generalization about self-organization of individuals explains the 50:50 distribution will be contingent on lower-level features within the individuals that permit individuals to modulate their behavior in response to behaviors of other individuals or external stimuli, by means of chemical or visual signals associated with the behavior of other individuals, as well as the genetic features that structure individual responsiveness.

What causes an individual to forage or rest at a given time? Here biologists have divided the types of causal contribution that are thought to interact in determining the behavior of an individual bee into external and internal components (Beshers and Fewell 2001). Both the external and the internal causal components are multiple. The external components include availability of resources (pollen, nectar, water), colony demography, how many brood are there, and how much pollen is already stored in the nest. These features of the environment are detectable by individual insects, and some of the mechanisms of their detection are beginning to be studied (Winston and Slessor 1998). For example, the number of brood is correlated with the concentration of a chemical pheromone that each offspring gives off. There is experimental work on the mechanism by which this chemical information modulates the foraging behavior of individuals by means of depressing the threshold for foraging for sucrose, and hence the more concentrated the chemical brood pheromone, the lower the threshold for foraging and the more bees will begin to forage at a given stimulus level (Page et al. 1998). The amount of pollen stored in the colony is thought to be represented to the individual foragers by means of a waiting time mechanism (Seeley 1995). Here the more cells in the hive that are already filled with pollen, the longer the time it will take for an unloading worker bee to find an empty cell in which to deposit the load and hence the longer the waiting time for the next bee to be unloaded. The longer an individual waits, the less likely it is to return to foraging. The shorter the time, the more likely an individual is to return to foraging.

These external factors modulate the behaviors of individuals, which then either aggregate or interact to determine the distribution of behaviors we call division of labor. It is clear already that the effect of the external factors is conditional on the internal state of the individual. Internal causes are some-times compressed into a description of the individual's motivational state. Thus, at this level, the causal question becomes, How likely is a particular in-dividual to forage for pollen given the external stimuli it confronts? Individual behavior is further decomposable into individual history, where learning is the mechanism which modulates individual behavior (Scheiner et al. 2001); sensory capacities, morphology, and developmental age (Calderone 1998); and characteristic threshold level (Pankiw et al. 2001). These individual fea-tures of a bee can also be considered to be the effects of the aggregation or interaction of factors at an even lower level of biological organization, namely, hormonal or cellular interactions, neuroanatomy, neurochemistry, and genes.

Conclusion

Often natural systems are subject to what we can analytically distinguish as separate, multiple causes. We study the causal structure of complex systems precisely by decomposing that complexity into component structures, mechanisms, or forces and studying the parts in order to understand the whole. The goal is to describe the contribution of each causal component independently and then proceed to determine the results of all the component forces acting simultaneously. This strategy is the basis for controlled experimental investigations where we try to construct two identical systems save for the variable whose causal contribution we are investigating. We can then vary the values of that causal component to determine the effects of it on the system.

This strategy works well in systems where the causal contributions of different variables are additive and represents the type of thinking behind strict formulations of causal laws. If nothing else is interfering, what is the effect of variable P *alone* on a system? If we discover that changes in P issue in changes in Q in systems otherwise identical, we believe we have discovered some important information about the causal structure of that system. Indeed, if we can serially implement this strategy for each of the contributing causes, then it is hoped that we can describe, explain, and predict the behavior of the system under study. The problem with evolved, complex systems is that the operation of a single causal factor is contingent on a host of different conditions. As I have suggested, it can be contingent on the operation of other factors in ways that change the very causal contribution it makes. A single factor can also be contingent on its location in a structure being constrained, amplified, or damped from factors either above or below. Thus, the decompositional strategy that has had much success in explicating causal dependencies cannot be straightforwardly applied to the types of situations I have described. Indeed, I have pointed out that controlled experiments may be misleading if they are the sole methods used to understand redundant systems.

One message of this discussion of evolved, complex systems and causal contingency is that the generalizations we can discover that determine the behaviors of such systems may be variably contingent, but nevertheless causally active. We should expect to find behaviors determined by no single factor, but by multiple factors. Complete decomposition is successful only if the effects of the components are additive. When additivity holds, then we can export the results of idealized experiments where we hold all other factors constant or randomize on them, determine the contribution of each single factor, and use that knowledge directly by adding it to the contributions of the other factors. But in general we should expect to discover causal dependence only through partial decomposition of interacting components and reconstruct the operation of multiple factors by methods of integration that go beyond simple addition.

It is not sufficient to say *that* generalizations in the special sciences are contingent and hence not lawful. Rather, one must detail *what kinds* of conditions they depend on and *how* that dependency works. Only by further articulating the variety of types of contingency that are found in multicomponent, multilevel, evolved systems can we accurately describe the differences in methodology and results of different scientific domains.

5.4. CETERIS PARIBUS: AN INADEQUATE REPRESENTATION FOR BIOLOGICAL CONTINGENCY

Introduction

Biological systems are evolved, multicomponent, multilevel, complex systems.[6] Their features are, in large part, historically contingent. Their behavior is the result of the interaction of many component parts that populate various levels of organization from gene to cell to organ to organism to social group. It is my view that the complexity of the systems studied by biology and other sciences has implications for the pursuit and representation of scientific knowledge about such systems. I argue that a proper understanding of the regularities in biological systems should influence our philosophical views on the nature of causal laws and, in particular, the role of ceteris paribus qualifications.

A well-known problem for the special sciences, and biology in particular, is the failure of generalizations about evolved, complex systems to meet what have been identified as the defining characteristics of scientific laws. This is alleged to be a serious problem because of the special role that laws play in science. They are what science supposedly seeks to discover. They are supposed to be the codifications of knowledge about the world that enable us to explain why what happens happens, that help us to predict what will happen in the future or in other circumstances, and that provide us with the tools to intervene in the world in order to reach our pragmatic goals. As such, they have been taken to be the gold standard of modern scientific practice. Philosophers have analyzed and reanalyzed the concept of a scientific law or

[6] This section was originally published as S. D. Mitchell, "Ceteris Paribus: An Inadequate Representation for Biological Contingency," *Erkenntnis* 57, no. 3 (2002): 329–350. The research was supported by the National Science Foundation, Science and Technology Studies Program, grant no. 0094395. It was improved by my participation in the "Working Group on Social Insects," organized by Robert E. Page, Joachim Erber, and Jennifer Fewell and sponsored by the Santa Fe Institute, and the serious discussion of earlier versions of the paper at the Max Planck Institut für Gesellshaftsforschung and at the Greater Philadelphia Philosophy Consortium. I thank Joel Smith, Jim Bogen, and Clark Glymour for helpful comments.

a law of nature in the hopes of specifying a set of necessary and sufficient conditions that postulations of laws have to meet in order to be the "real thing" and hence be able to perform the functions of explanation, prediction, and intervention. The "received view" of what conditions are required of a law include:

1. Logical contingency (having empirical content)
2. Universality (covering all space and time)
3. Truth (being exceptionless)
4. Natural necessity (not being accidental)

Some hold that laws are not just records of what happens in the universe but stronger claims about what must happen, albeit not logically, but physically in our world and hence have the power to dictate what will happen or what would have happened in circumstances that we have not in fact encountered. Thus, laws are said to support counterfactuals. It is not clear that anything that has been discovered in science meets the strictest requirements for being a law. However, if true, presumably Newton's laws of motion, the laws of thermodynamics, and the law of the conservation of mass-energy would count. The closest candidates for being a law and test cases for a philosophical account of scientific law live most commonly and comfortably in the realm of physics. Many philosophers have pointed out that few regularities in biology seem to meet the criteria for lawfulness enjoyed by the laws of physics.

How are we to think about the knowledge we have of biological systems that fail to be characterized in terms of universal, exceptionless, necessary truths? Their inferior status is sometimes blamed on the contingency of biological causal structures. The ways in which biological systems are organized have changed over time; the systems have evolved. Their causal structures thus not only could have been different but in fact were different in diverse periods in the evolution of life on the planet and in distinct regions of the earth, and most likely will be different in the future. Thus, exceptionless universality seems to be unattainable. The traditional account of scientific laws is out of reach for biology. Should we conclude that biology is lawless?

If so, how can we make sense of the fact that the patterns of behavior we see in a social insect colony or the patterns of genetic frequencies we see over time in a population subject to selection are caused, are predictable, are explainable, and can be used to reliably manipulate biological systems? The short answer is that biology has causal knowledge that performs the same epistemological and pragmatic tasks as strict laws without being universal, exceptionless truths, even though biological knowledge consists of contingent, domain-restricted truths. This alone raises the question of whether laws in the traditional sense should be taken as the gold standard against which to assess the success or failure of our attainment of scientific knowledge.

But perhaps we should not be too quick to abandon the standard. There is, after all, a well-worn strategy for converting domain-restricted, exception-ridden claims into universal truths, and that is by means of the addition of a ceteris paribus clause. Take the causal dependency described by Mendel's law of segregation. That law says: In all sexually reproducing organisms, during gamete formation each member of an allelic pair separates from the other member to form the genetic constitution of an individual gamete. So, there is a 50:50 ratio of alleles in the mass of the gametes. In fact, Mendel's law does not hold universally. We know two unruly facts about this causal structure. First, this rule applied only after the evolution of sexually reproducing organisms, an evolutionary event that, in some sense, need not have occurred. Second, some sexually reproducing organisms don't follow the rule because they experience meiotic drive, whereby gamete production is skewed to generate more of one allele of the pair during meiotic division. Does this mean than Mendel's law of segregation is not a "law"? We can say that, ceteris paribus, Mendel's law holds. We can begin to spell out the ceteris paribus clause: Provided that a system of sexual reproduction obtains, and meiotic drive does not occur, and other factors don't disrupt the mechanisms whereby gametes are produced, then gamete production will be 50:50. Finer specifications about possible interference, especially when they are not yet identified, get lumped into a single phrase – "ceteris paribus" – when all else is equal or provided nothing interferes. This logical maneuver can transform the strictly false universal claim of Mendel's law into a universally true, ceteris paribus law. With the ceteris paribus clause tacked on, even biological generalizations have the logical appearance of laws.

But the cost of the ceteris paribus clause is high. First, although making a generalization universally true in this way can always be done, it is at the risk of vacuity. Woodward (2002a) makes this argument clearly and rejects ceteris paribus laws entirely, advocating instead a revision of our account of explanation that does not require universality. Others, like Pietroski and Rey (1995) have suggested that there are ways to fill out the ceteris paribus clause to make it contentful. However, the ability to fully fill in the conditions that could possibly interfere may well be an impossible task. Indeed, in evolutionary systems new structures accompanied by new rules may appear in the future, and hence we could never fully specify the content of potential interfering factors. Still others, such as Lange (2000, 2002), have argued that vagueness is not equivalent to vacuity. Lange argues that scientists in their practice tacitly know what is meant by a ceteris paribus law. They know some cases of interfering factors and can extrapolate the nature of other factors by means of their family resemblance to the known ones. Earman, Roberts, and Smith (2002) maintain that there are strict laws to be found, at least for fundamental physics, so there is no need for ceteris paribus laws there. Furthermore, they argue that although the special sciences cannot discover strict laws, there are no such things as ceteris paribus laws. Their challenge leaves us with the problem of how to account for the explanatory and predictive power of biological generalizations if, as their account would entail, there are no laws in these domains. I argue that it is only by providing a detailed account of biological knowledge claims that we can hope to address the problem that Earman, Roberts, and Smith have posed.

Critics of the ceteris paribus clause correctly identify the fact that the clause violates the logical spirit of the concept of "law." I argue that, more importantly, it violates a pragmatic aspect of "laws" in that it collapses together interacting conditions of very different kinds. The *logical* cloak of ceteris paribus hides important differences in the *ontology* of different systems and the subsequent differences in epistemological practices. Whereas ceteris paribus is a component of the statement of a causal regularity, what it is intended to mark in the world is the *contingency* of the causal regularity on the presence and/or absence of features on which the operation of the regularity depends. Those contingencies are as important to good science as are the regularities that can be abstracted from distributions of their contextualized applications.

Indeed, a familiar way to mark the difference between exact and special sciences is by pointing to the contingency of the products of biological evolution, and the contingency of the causal dependencies to which they are subject. This is what Beatty dubs "the evolutionary contingency thesis" (Beatty 1997). Beatty explicitly attempts to give a rigorous account of Gould's metaphoric sense of contingency as rewinding and replaying the tape of life. In Gould's thought experiment, the second evolutionary sequence would reveal entirely different species, body plans, and phenotypes. The intuition is that small changes in initial "chance" conditions can have dramatic consequences downstream. Sexual reproduction itself is thought to be a historically contingent development, and hence the causal rules that govern gamete formation, for example, are themselves dependent on the contingent fact that the structures that obey those rules evolved in the first place. Biological contingency denotes the historical chanciness of evolved systems, the "frozen accidents" that populate our planet, the lack of necessity about it all.

There have been different responses to the mismatch between the strict, ideal version of scientific law and the products of the special sciences.

* Biology has NO LAWS on the standard account of laws (Beatty, Brandon). This may not be so bad, if, as Cartwright argues, science doesn't need strict laws or, as Woodward argues, we can have explanatory generalizations without universality.
* Biology HAS LAWS on the standard account of laws (Sober, Waters). There are not many and these are ceteris paribus laws or are very abstract, perhaps mathematical truths or laws of physics and chemistry.
* Biology HAS LAWS on a revised account. We need to reject the standard account of laws and replace it with a better account (Mitchell).

Some have opted to accept the lesser status of biological generalizations and preserve the language of law for those venerable truths that people such as Steven Weinberg dream will be few and so powerful as to make all the other knowledge claims we currently depend on part of their deductive closure. If laws are understood in the strict sense, biology doesn't have any, and the picture is even worse for the social sciences (Beatty 1995, 1997; Brandon 1997). Others, while accepting this conclusion, have gone on to suggest that laws are not so important, anyway, and so it is not so bad that the special sciences fail to have them; we don't need them (Cartwright 1994, 1999). Still others scramble to construe the most abstract of relationships within the special sciences or some physical or chemical regularity internal to biological systems as laws (Sober 1997; Waters 1998), thus there are some laws, though not many, and they are not clearly identified as distinctively biological (Brown and West 2000). There are some laws, nevertheless, and hence the legitimacy of the special sciences can be restored according to this line of argumentation.

I find these responses less than satisfying. We need to rethink the idea of a scientific law pragmatically or functionally, that is, in terms of what scientific laws let us do rather than in terms of some ideal of a law by which to judge the inadequacies of the more common (and very useful) truths. Woodward (2002a, 2002b) also adopts a strategy to reconsider the nature of laws in the special sciences, rather than force those claims uncomfortably into the standard view, wedged in with the help of ceteris paribus clauses. He had developed an account of explanation that requires generalizations less than universal in scope, but which can, nevertheless, support counterfactuals.

My strategy is somewhat different: I recommend that we look more closely at the character of the contingencies of the causal dependencies in biological systems that are often lumped into a single abstract concept of "contingency" and singled out as *the* culprit preventing biological science from being lawful

(Beatty 1997). I have argued elsewhere (1997, 2000) that general truths we discover about the world vary with respect to their degree of contingency on the conditions on which the relationships described depend. Indeed, it is true that most of the laws of fundamental physics are more generally applicable, that is, are more stable over changes in context, in space and time, than are the causal relations we discover that hold in the biological world. They are closer to the ideal causal connections that we choose to call "laws." Yet few of even these can escape the need for the ceterius paribus clause to render them logically true. Indeed, Cartwright has argued that we can't find a single instantiation of these laws, and Earman et al. (2002) admit that we have yet to identify (or at least have evidence that we have identified) a law of physics. What's going on here?

The difference between fundamental physics and the special sciences is *not* between a domain of laws and a domain of no laws. Yet I would agree that there are differences, and those differences can inform our understanding of not only the special sciences but the very notion of a "law" and its function. By broadening the conceptual space in which we can locate the truths discovered in the various scientific pursuits, we can better represent the nature of the actual differences. The interesting issue for biological knowledge is not so much whether it is or isn't just like knowledge of fundamental physics, but how to characterize the types of contingent, complex causal dependencies found in that domain.

Recent developments in understanding scientific causal claims take as their starting point less than ideal knowledge. They explore how to draw causal inferences from statistical data, as well as how to determine the extent of our knowledge and range of application from the practices of experimentation (Glymour 2001; Spirtes et al. 1993; Woodward 2000, 2001, 2002a). When we see that some relationship holds here and now for this population in this environment, what can we say about the next population in the next environment? These new approaches recognize the less-than-universal character of many causal structures. If explanatory generalizations were universal, then when we detect a system in which A is correlated with B, and we determine that this is a causal relationship, we could infer that A would cause B in every system. But we often cannot. The difficulty goes beyond the correlation-causation relationship.

In systems that depend on specific configurations of events and properties that may not obtain elsewhere and that include the interaction of multiple, weak causes rather than the domination of a single, determining force, which laws we can garner will have to have accompanying them much more information if we are to use that knowledge in new contexts.

Let me then consider the parts of this situation in turn. First, what kinds of complexity are present in biological systems? Second, what is the nature of the contingency of causal structures in biology? Third, what are the implications of complexity and contingency for scientific investigation?

Complexity

While the term "complexity" is widely invoked, what is meant by it varies enormously. Often linked with chaos and emergence, current definitions of complexity numbered somewhere between thirty and forty-five in 1996 (at least according to Horgan's report of Yorke's list; see Horgan 1996: 197, fn. 11). I believe the multiplicity of definitions of "complexity" in biology reflects the fact that biological systems are complex in a variety of ways:

- They display complexity of *structure*, the whole being formed of numerous parts in nonrandom organization (Simon 1981; Wimsatt 1986).
- They are complex in the *processes* by which they develop from single-celled origins to multicellular adults (Goodwin 1994; Goodwin and Saunders 1992; Raff 1996) and by which they evolve from single-celled ancestors to multicellular descendants (Bonner 1988; Buss 1987; Salthe 1993).
- The *domain* of alternative evolutionary solutions to adaptive problems defines a third form of complexity. This consists of the wide diversity of forms of life that have evolved despite facing similar adaptive challenges (Beatty 1995; Mitchell 1997, 2002c).

Compositional Complexity. Minimally, complex systems can be identified in contrast to simple objects by the feature of having parts. Simon defined a complex system as "made up of a large number of parts that interact in a nonsimple way ... the whole is more that the sum of the parts" (Simon 1981: 86). Complex systems are also characterized by the ways in which the parts are arranged, that is, the relations in which the components stand or their structure. The cells constituting a multicellular organism differentiate into cell types and growth fields (Raff 1996). A honeybee colony has a queen, and workers specialize in nursing, food storage, guarding, or foraging (Wilson 1971; Winston 1987). Such systems are bounded and have parts, and those parts differentially interact. In hierarchical organization the processes occurring at a specific level – for example, the level of an individual worker in a honeybee colony – may be constrained both from below, by means of its genetic makeup or hormonal state, and from above, by means of the demographic features of the colony. In Simon's terms, the system is partially decomposable. Interactions occur in modules and do not spread across all parts constituting the system. For example, in a honeybee colony if a need develops for more pollen, the workers whose task it is to forage for pollen will increase their activity, but changes need not occur in all other task groups (Page and Mitchell 1991).

Different types of organizational structures will mediate the causal relations within a level and between levels. Some modular structures shield the internal operations of a system from external influences, or at least from some set of them. Other features make structure the way in which external information is transmitted through the module.

Complex Dynamics. The nonlinearity of mathematical models that represent temporal and spatial processes has become, for some, the exclusive definition of complexity. Although this type of complexity is widespread, it still captures only one aspect of process complexity. Process complexity is linked with a number of dynamical properties, including extreme sensitivity to initial conditions (the butterfly effect) and self-organizing and recursive patterning (thermal convection patterns) (Nicolis and Prigogine 1989). Self-organization refers to processes by which global order emerges from the simple interactions of component parts in the absence of a preprogrammed blueprint.

The sensitivity of complex behaviors is further complicated by the fact that developing and evolving organisms and social groups are subject to the operation of multiple, weak, nonadditive forces. For example, in an evolving population natural selection may be operating simultaneously at different levels of organization: on gametes, on individuals with variant fitness relative to their shared environment, on kin groups of different degrees of relatedness, and so on (Brandon 1982; Hull 1989; Sober 1984c). In addition to multiple levels of selection, genetic drift may influence the patterns of change in the frequencies of genes through time. Phylogenetic constraints that limit the options for adaptive response, as well as physical constraints, such as the thermal regulation differentials for land-dwelling or sea-dwelling endotherms, may also stamp their character on the types of change available in an evolving population. One, some, or all of these different forces may operate in varying combinations.

Evolved Diversity of Populations. The third sense of complexity found in biology is exhibited by the diversity of organisms resulting from historical contingencies. Given the irreversible nature of the processes of evolution, the randomness with which mutations arise relative to those processes, and the modularity by which complex organisms are built from simpler ones, there exists in nature a multitude of ways to "solve" the problems of survival and reproduction. For example, a social insect colony adjusts the proportions of workers active in particular tasks in correspondence to both internal and external factors. If there is a loss of foragers, the younger individuals may leave their nursing or food-storing tasks to fill the vacant jobs. This homeostatic response is accomplished in different ways by different species. Honeybee colonies harbor sufficient genetic variability among the workers to generate variant responses to stimuli (Calderone and Page 1992; Page and Metcalf 1982). Ant colonies, on the other hand, may accomplish the same sort of response flexibility by means not of genetic variability but of variations in nest architecture (Tofts and Franks 1992). The ways in which information gets modulated through these systems depends on a host of properties. But what is important to notice here is that the manner by which one social insect colony solves a problem may well be different from solutions found by similar organisms.

Historical contingencies contribute to the particular ways in which organisms develop and evolve. History fashions which mutational raw materials are available for selection to act upon, which resources already present can be coopted for new functions, and which structures constrain evolutionary developments.

As argued above, complexity carries with it challenges to the scientific investigation of causal dependence and the discovery of explanatory laws. The discussion of redundancy and controlled experiments articulated those challenges. There is selection pressure to produce redundancy in critical biological processes. Yet, in confronting such systems experimentally, it is clear that simple control setups will be unable to detect the causal power of one of a set of redundant processes in a context when backup systems are operative. Only by using an ensemble of experimental designs, varying the contexts and the conditions, will redundant causes be tractable.

Which rules apply and when the rules apply is contingent. Complexity affords different kinds of contingency that must be understood to accommodate both the knowledge we have of complex systems and the practices scientists engage in to acquire this knowledge.

Contingency

It is not particularly useful to say *that* laws are contingent or that they can be rewritten as ceteris paribus generalizations without detailing *what* kinds of conditions they depend on and *how* that dependency works. Only by further articulating the differences rather than covering them over with a phrase denoting the existence of restrictions can the nature of complex systems be taken seriously. The problem of laws in the special sciences is not just a feature of our epistemological failings; it is a function of the nature of complexity displayed by the objects studied by the special sciences. Providing a more adequate understanding of laws in the special sciences requires a better taxonomy of contingency so that we can articulate the several ways in which laws are *not* "universal and exceptionless." In what follows I detail a taxonomy of different sorts of contingencies that can play a role in the operation of a causal mechanism. Knowing when and how causal processes depend on features of what we relegate to the context of their operation is central to using our understanding of causal dependence to explain, predict, and intervene. I have grouped the types of contingency by consideration of logic, spatiotemporal range, evolution, and complexity.

Logical Contingency. I have pointed out elsewhere (Mitchell 2000) the obvious point that there is a clear sense in which all truths about our world that are not logical or mathematical truths are contingent on the way our universe arose and evolved. Thus there is one type of contingency, that is, logical contingency, that applies to all scientific claims. The causal structure of our world is not logically necessary. This is as true for physical structures that might have been fixed in the first three minutes after the Big Bang, atomic structures that appeared as the elements were created in the evolution of stars, and the dependencies found in complex, evolved biological structures whose rules as self-replicating molecules changed with the subsequent evolution of single-celled existence, multicellularity, sexual reproduction, and social groups. All causal dependencies are contingent on some set of conditions that occur in this world, not in all possible worlds.

Space-Time Regional Contingency (No Sexually Reproducing Organisms, No Mendelian Law). There is another sense of contingency that is attributed to biological laws to distinguish them from the laws of physics that refers to restrictions in spatial and temporal distribution (see also Waters 1998). Mendel's law of segregation of gametes, for example, did not apply until the evolution of sexually reproducing organisms some 2.5 to 3.5 billion years ago (Bernstein et al. 1981). The earth is one or two billion years older, the universe itself ten billion years older. The conditions on which causal structures depend are not equally well distributed in space and time. Biological causal structures are certainly more recent than some physical structures and may be more ephemeral.

Evolutionary Contingencies: Weak Contingency. Even after the initial evolution of a structure and the associated causal dependencies that govern it, there may be changes in future environmental conditions that will break down those structures. With their demise, the causal dependencies describing them will no longer apply. This is what Beatty dubs "weak evolutionary contingency": that is, the set of conditions on which a particular rule depends for its maintenance by natural selection may change over time such that the formerly adaptive rule will no longer *be* the best biological solution to an environmental problem and will be replaced by something better. There are types of historical contingency, or restrictions on the domain of applicability of causal dependencies that attach more often to biological regularities because life arose much later than other material forms and may not hold on to its causal structures for as long. But universality never meant that the causal structure described in a law would occur at every point in space-time; rather, that whenever and wherever the conditions for a relationship between the properties did occur, interactions would occur according to the relation described by the law.

Strong Contingency (Multiple Outcomes Contingency). Beatty identifies a stronger sense of evolutionary contingency with "the fact that evolution can lead to different outcomes from the same starting point, even when the same selection pressures are operating" (Beatty 1995: 57). Here the focus is shifted from how likely or widespread the conditions are under which a causal relation will be operant to the uniqueness of the causal relationship or structure being evoked by those very conditions.

There are two ways in which strong contingency could occur. The first is when systems are indeterministic. If quantum processes have effects on macroscopic phenomena, then there can be cases where all the initial conditions are the same and the outcomes are nevertheless variant. Given mutations may be generated by radiation effects; this could well be a symptom of biological systems. The second is when deterministic systems are chaotic. Thus, there may be the "same" initial conditions, that is, the same insofar as we can determine them to be the same, that nevertheless give rise to widely divergent outcomes. Complex biological systems are paradigm cases of chaos.

For deterministic cases, the ceteris paribus strategy for forcing this type of irregularity into the form of a law has been invoked (see Sober 1997). The reasoning is that if systems are deterministic, there *must* be specific conditions, even if we can never know what they are or determine whether they have occurred, that will distinguish the causal structures of the divergent outcomes. Sober argued that those conditions that caused the selection of a given structure and its accompanying regularities, for example, the variations and fitness differences for the evolution of sexual reproduction and gametic segregation, could then form the antecedent of a strict law.

Beatty's own response to this type of maneuver is to suggest that there is no way to enumerate the conditions in the ceteris paribus antecedent clause to transform the biological generalizations that describe the causal structures of evolved biological systems. Earman et al. (2002) remind us that this is only an epistemological worry. It may be that we will be able to articulate the conditions that completely specify the numerous conditions on which some evolved structure depended and depends and hence be able to articulate the strict law governing that domain. If it is just a matter of knowing or not knowing all the conditions, then the very existence of ceteris paribus laws that cover evolutionary contingency is not called into question.

While it is clear how this strategy would work for weak contingency, how would it apply to strongly evolutionarily contingent regularities? Here, even when the causal conditions are operant, different and functionally equivalent outcomes are generated. If the relationship between the antecedent conditions and the evolved structure is probabilistic, then either one could invoke the ceteris paribus strategy and allow for strict probabilistic laws. If the multiple outcomes are the result of a deterministic, but chaotic, dynamics, then one could still claim that although we cannot discover the strict deterministic laws that direct the system into separate outcome states, they nevertheless exist. Once again, evolutionary contingent claims can be embedded via the specification of the contingent antecedent conditions in a strict law.

The problem I see with invoking the ceteris paribus response to Beatty's evolutionary contingency thesis is that it collapses the different ways in which a causal regularity can fail to be strict. By so doing it obscures, rather than illuminates, the nature of biological knowledge.

Complexity and Contingency

So far I have discussed the historical dimensions of chance and change characteristic of the evolved complexity of the domain of biological objects that make lawful behavior of the universal and exceptionless variety hard to come by. In addition, there are contingencies that confound the strict lawfulness of currently existing complex biological systems. These are the result of the multilevel compositional structure of complex systems and the multicomponent interactions at each level of those systems – complexity of both composition and process.

Multilevel Interactions. The operation of some causal mechanisms is contingent on the constraints imposed by their location within a multilevel system. For example, one could detail the causal relations describing the ovarian development in female bees, something that may well be a conserved trait from solitary ancestors to social descendents. The developmental laws that describe this process depend on both the appropriate genes and internal (to the individual bee) conditions for the triggering of the expression of those genes at certain stages in the process of cellular specialization. However, when individuals come to live in social groups, the context in which these internal processes have to operate may change. When a female honeybee develops in a colony in which there is a queen, then the worker's ovarian development is suppressed by means of pheromonal control from the queen. If the queen should be killed and the colony left without a queen, then the workers immediately begin to develop ovaries and produce haploid eggs. Thus the conditions on which a causal mechanism operates depends on the organizational structure in which it is embedded. This is not an incident peculiar to social groups.

The story of ovarian suppression in social insects is similar to the change of rules that occurs when the single-celled stage of the slime mold ends with the aggregation to form the multicellular flowering stage, discussed above. Indeed, Buss (1987) has argued that the very origin of multicellular individuals from single-celled ancestors is one that involves the suppression of competition among the components (cells or worker bees) of the new individual (multicellular individual or colony), which permits the new individual to become the stable locus of evolutionary change. Thus, the composition rules that characterize the various ways in which complex biological objects are formed will also affect the nature of the contingency in which the component mechanisms will operate.

Multicomponent Causal Structures within a Level. In addition to the type of contingencies arising with compositional hierarchies, there are also contingencies that affect specific causal mechanisms occurring within a level of organization. These are the result of the fact that most behaviors of complex biological systems are the result of the interaction of multiple mechanisms or causal factors. That there is more than one force acting to produce an outcome does not, by itself, threaten the existence, empirical accessibility, or usefulness of strict laws describing the individual components. Rather, the problem arises with the nature of the interaction of these components. Positive and negative feedback loops and amplifying and damping interactions of a nonlinear type are characteristic of complex biological systems.

For example, current research indicates that the behavior of individual foragers in a honeybee colony varies with genetic differences. This has been described in terms of the threshold levels for the stimulus required to initiate foraging behavior of individual bees. Experiments have supported the view that genetic variation accounts for behavioral variation via the genetic components determining individual threshold levels. However, it is now known that learning from the environment can also affect the behavior of foragers by moderating their threshold level. Indeed, the contribution of each of these very different mechanisms can amplify the expected probability of foraging behavior. Thus, there may not be a "regular" manner in which the contributions of different components generate a resultant outcome. Under some values of the components, it may be that the contribution of genetic variation is much stronger than variant learning experiences and hence completely determines the pattern of foragers in a colony or between colonies. Other times, the reverse might be true, and still other times, the interaction of the two components is operative in generating the pattern. Interactions may take different forms, including adding, amplifying, swamping, and damping. Thus, the operation of a single causal component, its contribution to the resultant effect, can be contingent not just on background standing conditions, but also on the other causal mechanisms operating at the same time. The nature of the contingency may vary with the different values that the variables in the component mechanisms take.

Redundancy and phase change phenomena can also be present in complex biological systems (see section on "Redundancy and Phase Change" above). These offer two more types of complex contingencies that have import for understanding the range of contexts in which the regular behavior of a set of variables may be disrupted. In systems with redundant processes, the contribution of any one may be elicited or moderated by the operation of another. For phase change, like nonlinear dynamical processes, the nature of the function describing the behavior of variables may itself change under certain values of the variables or changes in external conditions.

The Philosophical Consequences

I have presented a variety of different ways in which causal dependencies in biology are contingent. This is damning of biology only if one retains the strict notion that laws must be universal and exceptionless. Instead, we can turn the question around and ask not whether biological claims can be transformed into strict laws but, rather, when and how do biological claims perform the functions that laws are thought to serve? That is, how can less than strict laws explain, predict, and assist in intervention? Recent work on causal dependence has done much to develop answers.

Woodward (2000, 2001, 2002b) applies his account of explanation being grounded not by universal, exceptionless laws but by generalizations that are invariant under intervention to cases in biology. He argues that his notion of invariance is distinct from my idea of stability of conditions on which relations hold, and that invariance is what is needed for explanation, not stability. I want to explore Woodward's argument, to see where and why the disagreement occurs. First, there are many similarities in Woodward's and my account of laws. We both reject the universality and exceptionlessness as necessary for knowledge claims to be deemed laws. We both want laws to have properties that come in degrees, rather than dichotomous values. What Woodward identifies as the relevant continuum is that of invariance: "...unlike lawfulness, invariance comes in gradations or degrees" (2000: 199). For him, generalizations come with different degrees of invariance. It seems a bit odd to say invariance comes in degrees, since it seems to be the case that the relationship between two variables is either invariant or not. To say that how much it varies and how much it fails to vary can be tracked doesn't seem to make linguistic sense. However, Woodward does make sense. He means that the relation between two variables is domain-insensitive, such that if one changes the value of the independent variable X, the dependent variable, Y, will stand in the same functional relationship, say, $Y = -kX + a$ for a range of changing values of X. There are, however, some values that X can take for which the function no longer is true. So the function is invariant under some but not all changes in the value of X. The number of changes, or how large the domain of invariance is, of different functions will differ, and this is where I believe the degrees come in. Some functional relationships hold universally, some hold nearly universally – except, say, near a black hole – some hold for the majority of the time, and some hold some of the time. The value of X thus explains the value of Y by means of the functional relationship that describes the causal dependence of Y on X, in just those regions of the domain where the relationship is invariant. And for Woodward, having some domain of invariance is sufficient for explanation, even if it isn't universal, since some counterfactual situations – namely, those changes in X where the function remains invariant – are supported.

I also attempt to provide a means of describing varying domains of applicability of different scientific laws (Mitchell 2000). To this end, I have identified a number of different continua in which generalizations can be located – in particular, ontological ones of stability and of strength and representational ones of abstraction and cognitive accessibility. Ontological differences obtain between Fourier's law of thermal expansion, for example, and Mendel's law of segregation, but it is not the case that one functions as a law and the other does not, or that one is necessary and the other is contingent. Rather, one difference is in the stability of the conditions on which the relations are contingent. Once the distribution of matter in the primordial atom was fixed, presumably shortly after the Big Bang, the function described by Fourier's law would hold. It would not have applied if that distribution had been different and, indeed, will not apply should the universe enter a state of heat death. The conditions that both gave rise to the evolved structure of sexual reproduction and meiotic process of gamete production are less stable. The strength of a deterministic, a probabilistic, and a chaotic causal relation also varies.

There are methodological consequences to these variations in stability and strength. There is a difference in the kind of information required in order to *use* the different claims. It would be great if we could always detach the relation discovered from its evidential context and be assured it will apply to all regions of space-time and in all contexts. But we cannot. Causal structures are contingent, and as I have argued above, they are contingent in a number of different ways. To apply less than ideally universal laws, one must carry the evidence from the discovery and confirmation contexts along to the new situations. As the conditions required become less stable, more information is required for application. Thus, the difference between the laws of physics, the laws of biology, and the so-called accidental generalizations is better rendered as degrees of stability of conditions on which the relations described depend, and the practical upshot is a corresponding difference in the way in which evidence for their acceptance must be treated in their further application.

Woodward compares his idea of invariance with my idea of stability and finds mine wanting for the job of explanation. Stability for me is a measure of the range of conditions that are required for the relationship described by the law to hold, which I take to include the domain of Woodward's invariance. However, stability can be a feature of relationships that are not invariant under ideal intervention. "Mere stability under some or even many changes is not sufficient for explanatoriness" (Woodward 2001). His counterexample is the case of common cause. According to Woodward, while the relationship between the two effects of a common cause is stable in any situation in which the common cause is operative, the one effect does not explain the other effect. Woodward's notion of invariance is supposed to capture this distinction, since some ideal interventions in the common cause system will show that a change in the value of the first effect will not be correlated with a change in the value of the second effect. The relationship breaks down and so is not explanatory. If the world were such that those types of interventions never occurred naturally nor could be produced experimentally, on my view stability would be maintained, but invariance would still be transgressed, since there could be ideal situations in which it broke down. So, on his view, we would not have a "law"; on mine, we would.

The empiricist in me finds it difficult to detect the cash value of the difference Woodward is drawing between invariance and stability. If we could produce or witness the breaking of the relationship between the effects of a common cause, then we would find that the law describing the relationship between cause and either effect to be more useful than a law describing the relationship between effect and effect. Namely, the former would work in cases where the latter did not. But if there never were such cases to be found, then wouldn't they work equally well for prediction and intervention? If, on the other hand, one requires a more substantial metaphysical warrant for explanation, as I believe Woodward does, then this constant conjunction or stable correlation would fail to explain why what happens happens. The trouble is, I do not see how in practice one can distinguish the positions. If there is evidence for a common causal structure, then on both accounts, the cause is a better predictor and better at explaining the effects than either effect is of each other. If there is no evidence for a common causal structure, then the correlation, with the right sort of supporting evidence (such as temporal order), would be taken to be explanatory.

I think the disagreement lies in the functions of laws on which we individually focus. Woodward admittedly attempts to account for only the explanatory function of laws. To perform this function, a generalization must report a counterfactual dependence. It has to describe a causal relationship that will remain true under certain episodes of "other things being different." It need not be true under *all* such episodes, that is, be exceptionless and universal, but only to explain a particular occurrence. Thus, to say why a variable Y has the value it has by appeal to the value of a variable X, it must be the case that one could track x, y pairs in other circumstances, in particular, in interventions, *in the domain where the law was invariant,* and find the same relationship one finds in the explanandum situation. This is for Woodward what it means to say that Y is causally dependent on X, and hence an occurrence or value of it is explained by appeal to X by means of the invariant generalization connecting X and Y. That is, Woodward lets domain-restricted generalizations count as explanatory in just those domains where the relationship described in the generalization holds. Stability does just the same work; however, it is weaker and includes what might turn out to be correlations due to a nondirect causal relationship. But for there to be a distinction between stability and invariance, then we would have to already know the causal structure producing the correlation.

Because the sciences I worry about embrace complexity, my goal has been to see how complexity affects the way we do science. Now if the world were hopelessly complex or the dynamical evolution so rapid that we found ourselves in a Heraclitean universe, we wouldn't be able to capture any knowledge that could be used downstream. But the history of science doesn't make this a plausible view – we can't be that deluded for that long that we actually can manipulate and predict events in the world. But it is equally misguided to take as an assumption that the world is simple, and expect to find that simplicity at every turn, and blame the investigator or impugn the science when simple laws are not to be found. We can learn about the features of a complex world; it is just not easy, and no single algorithm is likely to work in all the contexts in which complexity is found.

So where does that leave us with respect to the implications of complexity and contingency for our epistemological practices? First, for descriptions of causal dependencies to be useful for prediction, explanation, and intervention, they need not be universally true and exceptionless. As long as we can detail the domains in which the dependency is stable or invariant, then we can explain why what happens in that domain happens, and what will happen when there are changes in the magnitude of the causal parameters. However, there are different ways in which domains are restricted, or universality is lost, including temporal and spatial restrictions that are the result of the evolutionary process; contextual restrictions in which certain parameter values or background conditions change the functions that describe the causal dependency; and contextual restrictions in which the operation of other causal mechanisms can interact in ways in which the effects of a cause are evoked, amplified, damped, or made redundant.

Representing all that variety of contingency by means of an unspecified ceteris paribus clause will mask the different strategies required to elicit information about complex contingencies in nature. In short, the context-sensitivity of complex dynamical systems, like those studied by biology, implies a shift in our expectations. We should not be looking for single, simple causes. We should not be looking exclusively for universal causal relationships. And we must record and use not only the causal dependencies detected in a particular system or population to understand other systems and populations, but also the features that define the contexts present in the system under study. Without that information, exporting domain-specific, exception-ridden general truths cannot be done.

6

Pluralism or Disunity

The argument of this chapter is that the complexity of biological objects generates a plurality of theories, models, and explanations. With this antireductionist picture of science come new questions. In what relationship do these multiple accounts stand to one another or to the objects they describe? If there is not one, or a few, laws of biology from which all else can be explained, then how can we judge which of the infinite possible generalizations or explanations should be accepted and which not? If there is a pluralism to be had, what kind of pluralism is it?

In this chapter I first consider why reduction is attractive as an account of the relationship among scientific theories. I then argue that it fails, and pluralism is a better model. I consider two defenses of pluralism, one by John Dupré (1983, 1993, 1996), the other by Paul Sherman (1988, 1989) in the spirit of Niko Tinbergen, and reject both of these as inadequate. In the first, Dupré argues for a degree of promiscuity of taxonomies and theories that seems to me to be too unrestrained. His defense of pluralism, or disunity as he calls it, is intriguing but, I argue, ultimately fails. I propose a different argument to support a picture of pluralism that is sympathetic to, if not coextensive with, the one suggested by Dupré.

The second source of pluralism for biology, in particular, harkens back to Tinbergen's four-questions description of the conceptual map of biological science. He suggested that there are distinct questions one might raise of the same phenomenon, and hence that different, multiple answers are to be expected. Paul Sherman revived this partitioning of biological theories and explanations in order to sort out disputes between developmental and selectionist explanations of current biological traits. I argue in the last two sections that this picture of pluralistic biology mislocates the arenas for disputes. Instead, a more nuanced understanding of the scientific representations of multicomponent, multilevel, evolved complex systems permits a more accurate

179

understanding of the integration of the plurality of models called up in biological discourse. In the last section I put this argument to the test, by evaluating the relationship between different self-organization models for division of labor in social insects.

6.1. CRITICS OF UNITY OF SCIENCE

Philosophers and scientists continue to debate the unity or disunity of science.[1] In particular, sides are taken regarding the type of unity forged by the reduction of the "special sciences" to physics versus the disunity resulting from the autonomy of those sciences. When the suggestion is made that the theories and methods characterizing one scientific discipline, for example, biology, have contributions to make to the solution of problems found in separate disciplines, for example, psychology or sociology, the fear of expropriation arises. There appears to be a developing antireductionist consensus, especially among philosophers of biology. The consensus is that a simple derivability relationship between accounts of macroscopic phenomena (organisms, minds, societies) and accounts of microscopic phenomena (cells, molecules, atoms) of the kind long advocated by many philosophers of science is inadequate to capture the rich variety of relations among the results of scientific inquiry.

The defense of a disunified, pluralistic picture of science celebrates diversity and variety. Indeed, this picture is sometimes aligned with noble political sentiments such as democracy, freedom from patriarchy or racism, and rejection of narrow hegemonic social values. But politics is fickle when it pairs up with images of science. Indeed, Peter Galison has recently suggested that in mid-nineteenth-century Germany and then between the wars in the twentieth century, unity of science was associated with the call for freedom from nationalism and fascism. Yet in 1830, Comte's first lesson, "One cannot reduce all the sciences to a unity," was offered in an effort to protect science from appeal to nonscientific universal explanations. Galison judges, "What we will not find is a single-valued transhistorical function that plots assessment of unity onto a fixed political map" (Galison 1996: 8). Thus, while political arguments are well and good, one can't infer from them how science should look. Instead, you have to look at science directly.

---

[1] This section has not been published previously.

Why Reductionism Is Compelling

A compelling argument can be made for reductionism. We live in one world, not many worlds. Further, the material from which all the entities in the world are built is ultimately one kind of "stuff," that is, matter. It is the job of scientists to represent the features of that one world in a way that allows us to explain the patterns of phenomena we observe, to predict what will occur in future situations, and to permit us to intervene in ways that allow us to accomplish our practical goals. The next step in this argument is to add the assumption that the representations scientists come to accept as true stand in some sort of strong mapping relation with the actual features or structures of the world. The strongest form of this argument would assume that the representations most widely accepted by the scientific community are direct mirrors of the world's structure.

If this were the case, then we would expect generally accepted but diverse representations of a given domain, for example, the different accounts of division of labor in social insects, to stand in some strong mapping relation to each other. What is that relation? Minimally, we would expect the representations to be consistent with each other. If the world is one, and scientific claims accurately describe it, then two contradictory statements cannot both be true. But reductionists argue for a stronger relation than this. They argue for intertranslatability, or derivability. Ultimately, there is one fundamental, maximally accurate description from which the others can be derived. These are the earmarks of a reductionist view of science. While intertranslatability is symmetric, derivation is not. There is a preferred ordering to the direction of such a reduction, and that ordering is generally provided by another metaphysical assumption, namely, compositional materialism.

The material composition assumption that every object is made up of one type of substance, namely, matter, suggests that there is some basic level of description of the material building blocks (Moser and Trout 1995). The privileged level of description is taken to be the most fundamental, and one can come to understand the more complex objects by knowing the properties of the simple components and the composition functions. So if atoms make up molecules, and molecules constitute chemical elements, and elements make up different types of material objects, and material objects are the parts of cells, and cells make up organisms, and organisms make up societies, then if we understand atoms (or quarks or whatever we take as fundamental) and how these combine to form less fundamental objects, that should suffice. Scientific truths of biology then could be restated as truths about chemistry, those in chemistry in terms of physics. So the reasoning goes.

Indeed, if we believe that a description at the most fundamental level – physical or material description – is sufficient to describe the causal interactions responsible for all changes of state, that is, if we endorse the doctrine of causal completeness, then while descriptions, explanations, and predictions in a language of biology might be convenient (and true, if translatable or reducible to the fundamental level of representations), they are not necessary. We could, in principle, reduce all the diversity of the representations of contemporary science to statements describing the fundamental elements (see Causey 1977).

Why Reductionism Doesn't Capture the Realities of Scientific Inquiry

Grounds for rejecting reductionism for all cases are found in a more comprehensive analysis of the nature of scientific representations. The required simple mirroring relationship between theory and world does not hold. Since Kant, most philosophers accept that every representation will be shaped, in part, by the concepts that humans bring to the task of describing the world. While compositional materialism may be correct that all entities are made up of matter, the inference to a logical "composition" relationship between the "entities" in our scientific theories is not immediate.

Scientific representations are abstractions or idealizations (Cartwright 1980, 1982, 1989, 1994; Dupré 1983, 1993, 1996; Wimsatt 1987). They can represent only partial features of individuals rather than the individuals themselves as complex causal agents. For example, parasite and host theories explain population-level interactions and identify abstract individuals in terms of their roles in those interactions. The theories are then used to explain actual interactions between concrete individuals such as particular malaria viruses and humans. Yet the abstract individual identified by its functional role in the population models clearly does not exhaustively describe the individual as a causal agent (Dupré 1993). An individual human being is truly described in different theories at the same time as a host to a parasite, a consumer in an ecosystem, and a phenotypic expression of a set of genotypes, as well as a mammalian organism, a homeostatic endotherm, an organization of multiple cell types, and so on.

Actual, complex events or concrete individuals – the constituents of the one-world ontology – are, at the same time, instances of objects of multiple abstract theories concerned with different compositional levels. Reductionism requires replacing the higher-level abstractions by lower-level ones. Yet the abstractions, which constitute theoretical objects at the different levels, do not constitute identical representations across levels. That is, even if the descriptions at the various levels are all accurate, by being partial they may not be representing the *same* features of nature and hence would not stand in any straightforward derivability or intertranslatability relation nor form a neat, nested hierarchy. Dupré defends this type of picture. He is a materialist but denies that "there is any interesting sense in which ontological priority must be accorded to the allegedly homogenous stuff out of which bigger things are made" (1993: 89). Thus, Dupré endorses a weak, compositional materialism, that is, the view that "whatever kinds of things there may be, they are all made of physical entities" (92) that permits antireductionism. The material from which entities are composed does not carry all the explanatory weight. However, it is the association of matter with cause and the further alignment of cause with explanation that makes materialism and antireduction difficult bedfellows. In considering these issues in general, Dupré addresses the argument that runs from the requirement that "explanations at different structural levels must at least be consistent with one another." That such consistency is ensured by reduction is obvious, but, Dupré points out, the only way to reach such consistency entails that reduction rest on a particular view about causality, namely, the assumption of causal completeness.

Thus, reductionists present an even stronger argument that higher-level descriptions are dispensable. Causal closure is a doctrine about sufficient cause. Dupré considers this argument and rejects it. Dupré's reconstruction of the reductionist's argument is as follows: Consider an event at the microphysical level. Dupré uses the example of the motion of an electron in one's index finger. As a person moves one's finger to type the letter "b" on a keyboard, the electron has to move. If the microlevel is causally complete, then there is a set of events at that level, prior to the motion of the electron, that is sufficient to cause it to so move. Nevertheless, an explanation might be given at a macro level, that is, by appealing to one's intention to type the letter "b" to explain the motion of the finger that, of course, entails the movement of the electron. Since what is going on at the micro level, given the assumption of causal completeness, is sufficient to bring about the motion of the electron, the macro story must be at least consistent with the micro causal story. Furthermore, since causal completeness presumes sufficiency at the micro level, nothing more at the macro level is necessary to bring about the movement of the electron. Hence, macro-level events appear to be causally inert. Since on this account explanation is in terms of causes and all one needs for the causal story is micro-level events, all explanation could be run at the micro level. But this is just to restate the doctrine of in-principle reductionism.

This is a powerful argument, and one that sits behind many contemporary defenses of some form of reduction. In the face of it, many have promoted what Dupré correctly identifies as a weak form of reductionism, namely, supervenience. Others have bitten the bullet on causal explanations being the domain of the micro level and have opted for a defense of some other kind of explanation as appropriate to macro-level science (see MacDonald 1986). But Dupré wants to defend a much more robust pluralism and so claims that "... a central purpose of the ontological pluralism [he has] been defending is to imply that there are genuinely causal entities at many different levels of organization. And this is enough to show that causal completeness at one particular level is wholly incredible" (1993: 101). That may well be his purpose, but we have to look at whether or not he can get there with the arguments he provides. Dupré attempts to invert the reductionist *modus ponens* (causal completeness requires reductionism) into an antireductionst *modus tollens* (the failure of reductionism implies the falsity of causal completeness). While Dupré's account of the relationship between the two theses of causal completeness and reduction is correct on a purely ontological interpretation, his arguments do not accomplish the inversion he desires.

What Dupré uses as evidence that reductionism fails is the implausibility of it succeeding in any of a variety of the cases in biology and psychology that he considers. But all this shows is that reduction is unavailable *in fact* in these cases. It does not show that it is impossible *in principle*. It is the second, stronger claim that is needed to overturn causal completeness, for it is *in principle* reduction that figures as the conclusion of the reductionist argument.

I want to suggest a different argument that can be launched against reductionism. The reductionist argues that *if* the world is composed of physical matter and *if* all composite entities are made up of just physical matter, then *if* there is an account of the complete cause of events at the physical level, nothing more causally, and hence explanatorily, is required by higher-level descriptions. But how do we understand that? There are two ways.

First, metaphysically, it is the view that there is at the physical level a single, unique process of physical interactions that bring about a physical result. Unless one adopts a position of uncaused events, this does seem inescapable. But what about representing this process in physics? One can believe that all events are caused and at the same time argue convincingly that all the factors contributing to the complete cause cannot be represented by any single theory (and the representational mechanism that instantiates it) in physics. The local, contingent constituents of every causal process just are not included in the scope of physical theory, as Cartwright has argued (1994), and these will always be part of the complete cause. So there may well be causal closure at the level of physical entities, while there is always incompleteness, or causal openness, in the representation or theorizing about those processes, the representations that make up the *physics* entities.

There is a further source for disarming causal completeness as the ground for reduction. That is the view that there are two aspects to composite, complex objects or events: the material and the manner in which the material is arranged, that is, matter and form, or material and structure. Some patterns in the world that we identify as causal processes may depend as much on the structural characteristics of complex objects as on the material characteristics. In fact, as Bechtel and Richardson (1993) have suggested, there may well be a continuum of contributions from matter and structure such that actual causal processes occupy many different locations.

How does this bear on causal closure? Well, if the physical level is construed only materially, then structure is a level up and causally significant. Hence causal closure is false. If, however, structure is included in the physical level, then macro objects are physical and closure applies to a collection of micro and macro objects and events. Either way, reduction to a purely material physical level is thwarted.

By breaking the connection between physical material and physics representations, the arguments for causal completeness and theory reduction are no longer as closely linked, and Dupré's reverse inference from nonreduction to noncompleteness does not go through. Nevertheless, there are conceptual and empirical arguments against the inference from causal closure to reduction that I believe make a strong case for the rejection of reductionism.

There are recent, seemingly new, arguments for reductionism, especially in the "new wave" reduction of psychology to neurophysiology (Bickle 1997; Churchland 1986). These permit appropriate modification of terms and claims of the two sciences to accommodate the elimination of false claims by reduction to lower-level true claims, thus recognizing that the mapping between terms need not be direct. There are specific objections, concerning the translatability, eliminability, and causal efficacy of functional properties that challenge the newer versions of reduction (Schouten and De Jong 1999). Each new wrinkle in the reductionist position elicits new responses from antireductionists (Chalmers's annotated bibliography lists eighty-two entries as of 2002 under "reduction," most of them in since 1980; see http://www.u.arizona.edu/~chalmers/biblio.html). Rather than counter every new twist in the reductionist argument, I instead attempt to articulate and explore an alternative view of the relations among scientific claims.

Alternatives to Reductionism: Types of Pluralism

Anything Goes. "Anything goes" pluralism is associated with Paul Feyerabend's epistemological anarchism (Feyerabend 1975, 1981). In the absence of foundational standards of justification, Feyerabend argues that any and all forms of argument are acceptable in science. Some versions of social constructivism (see Collins 1982, 1985; Latour 1988; Woolgar 1988) can be taken to promote "anything goes" pluralism, at least as far as truth is concerned. Perhaps it would be more aptly identified as "anything that scientists accept goes." This shifts the criteria by which different scientific claims are granted authority from their representational adequacy to their social authority. Thus, one might argue that as long as some group of self-identified scientists agree to accept a claim about the world, then it "goes" and the constraints on the ways in which scientific groups are formed may be loose enough to condone an interpretation of "anything goes."

I find *both* the advocacy of retaining all, possibly inconsistent, theories that emerge from a community of investigators *and* the insistence that any collection of analyses of the same phenomena must be reduced to a single theory equally unacceptable. The challenge is to define clearly the middle ground: How can a diverse, well-confirmed, but irreducible set of theories be used collectively to achieve a more complete understanding than any of the theories taken in isolation?

Isolation of Theories. A prima facie problem for reductionism is the apparent diversity of theories in a science such as ecology. It has been argued that some of the compatibility of diverse accounts can be explained by the divergence of questions and interests that the scientist brings to the table. Mayr, Tinbergen, and Sherman have all defended a "levels of analysis" account of the plurality of theories and explanations that is almost dogma within the biological community (Mayr 1961, 1982; Sherman 1988; Tinbergen 1963). Mayr distinguished between proximate "how" and ultimate "why" questions in the face of the possible encroachment of genetics into whole organism biology (Beatty 1994). Tinbergen elaborated four different kinds of questions posed in the study of ethology. A recent revival of this model by Sherman further subdivides the four "levels" of questions that partition biological research. These are: evolutionary origins, functional consequences (the two "why" questions), ontogenetic processes, and mechanisms (the two "how" questions). Sherman's addendum divides questions of mechanisms into those that target physiology and those that target cognition (Sherman 1988).

For example, biologists explaining division of labor in social insects might approach this phenomenon with four different questions. Those concerned with evolutionary origins would investigate why this behavior arose when it did multiple times in the lineages of social insects; those with functional concerns would ask what current consequences the trait has on colony ergonomic efficiency and hence fitness; those with queries about ontogenetic processes would aim to explain how the various behaviors come to be expressed in the individual insects over time; and those interested in mechanisms would detail the environmental triggers and hormonal or cognitive mechanisms that then issue in the behaviors. Different questions invoke different explanatory schemata. Sherman claims, "Every hypothesis in biology is subsumed within this framework; competition between alternatives appropriately occurs within and not among levels" (Sherman 1988: 616). Thus, answers to questions at the different levels represent compatible components of a pluralistic, multi-dimensional body of knowledge. On this view, there is no need to account for intertheory relations among the levels.

Even if we granted that this describes current scientific practice, the question remains as to what relations *should* connect these autonomous enterprises. One reading of this type of compatibilism leads to an isolationist stance with respect to the separate analyses. If there is no competition between levels, there need be no interaction among scientists working at different levels either. The problem with the isolationist picture of compatible pluralism is that it presupposes explanatory closure *within each "level of analysis"* and a narrowness in scope of scientific investigation that precludes the type of fruitful interactions between disciplines and subdisciplines that has characterized much of the history of science (see Darden and Maull 1977). Even when the unlikely situation occurs that a scientist is narrowly concerned with only one level of analysis, it is a mistake to think that the answers to other questions have no bearing on the investigation at that level. There may be causal dependence or causal interaction between processes described by the different analyses. If so, remaining within a single level will fail to provide understanding for the questions addressed *at that very level* (Bechtel and Richardson 1993). Thus, the division of levels of analyses, different questions with different answers, could be mistakenly interpreted as a justification for unproductive heuristics for scientific investigation.

The answers to different questions appeal to distinct abstract models of causal processes as well as specify their application to concrete instances. Fundamental scientific laws used in explanations of concrete phenomena have ceteris paribus clauses (Cartwright 1980). In other words, they describe what is to be expected in idealized situations, when only one of a set of potential causal factors is operating, that is, when nothing else is interfering. All simple models also suffer by design from this type of unrealism. If we accept that multiple causal factors can, and often do, interact in the production and maintenance of human cultural practices, that the real cases are complex, without the ceteris paribus proviso, the laws would be literally false. With the proviso, however, such models do not themselves directly account for many, if any, real cases. That is, while abstract models describe the effects of the operation of a single causal process, our world does not normally approximate the ideal world that the models directly represent. The concrete explanatory situations on which we bring the abstract models to bear are messy, perhaps unique products of historical contingencies and interacting, multiple causal factors. In this sense, the simplified, abstract, theoretical models developed in science are literally false (cf. Wimsatt 1987). This does not make them undesirable or useless (cf. Richerson and Boyd 1987). Indeed, simplification allows models to be mathematically and empirically tractable for increasing precision of assumptions, crisper testability, and allowing counterintuitive results to be generated. The robust convergence of results of a variety of simple models is evidence that the result does capture a feature found in the complex world.

Nevertheless, the pragmatic virtue of simplicity is most frequently bought at the cost of realism in explanation (Levins 1968). Though, of course, it is possible that a single causal process may completely determine a particular event, given the complexity and diversity of human cultural phenomena, simple models are more likely to capture partial causes. In this situation, individually more complete models may be developed, if there are means for replacing the false assumptions in the model by more realistic descriptions (say, in models of language acquisition or measures of genetic fitness). At the most concrete level in generating an explanation, a model may introduce all of the relevant features that uniquely characterize a given event (say, the ecological conditions of a particular lake). However, the cost here is that of generality. A full-scale map of a town would express the greatest realism; however, it would be as useless for finding city hall on a map as one that represented the town as a single point. So, too, all that is true in a concrete model of Clear Lake will likely not apply directly to a neighboring lake. It is by distinguishing the idealized models from their applications that we can identify the location and scope of integration.

In short, isolationist pluralism employs a levels-of-analysis framework to endorse a strategy of limiting interactions between various theories offering explanations in a given domain. While some scientists may restrict their interests to a specific level only, this is not necessarily the case. Pluralism better describes the causal models that, by modeling the contribution of individual causes, necessarily abstract away the operation of other compounding factors. By so doing, they can make no incompatible claim about the operation of the ignored causes. Once this structure of causal models is recognized, one may understand why competitive interactions arise within a level (e.g., how best to measure fitness in determining the current function of a cultural trait). However, the model must be distinguished from its application. In application, one can immediately see that causal models that provide answers at different levels are indeed related. Thus, although pluralism is to be defended, it is not the pluralism of questions and the consequent independence of answers, but rather a pluralism of models of causal processes that may describe contributing factors in a given explanatory situation. This is not to recommend an "anything goes" pluralism. Not all explanations are equally good. Hence, to defend a strategy of pluralism for causal models and criticism of explanatory applications of those models requires a further account of how idealized models are to be integrated into explaining concrete, nonideal cases.

Integration. I have suggested that there are two broad dimensions to the plurality of scientific theories and explanations: a vertical arrangement of accounts of compositionally related objects of investigation and a horizontal relationship of different questions addressed to the one and same phenomenon. I have argued that the relationship among accounts across the compositional dimension is not one of reduction. In addition, the relationship among the accounts across the levels of analysis should not be one of isolation. I hold that the remaining plausible candidate for characterizing these relationships is a set of various forms of nonreductive integration.

I have argued elsewhere that the dual complexity of the phenomena studied by scientists and the diverse interests and pragmatic constraints on the representations scientists devise to explain the phenomena conspire against simple pictures of scientific knowledge (Mitchell 1997, 2000). Correspondingly, the strategy for integrating diverse theories and explanations will not be algorithmic. This is evident from even a superficial investigation of how genetics and population biology are jointly modeled, or of current models of how the biochemistry of hormone production in a developing organism affects and is affected by the external environmental conditions in which the organism finds itself. The genetics of a population will constrain the variation on which natural selection can operate, and the operation of natural selection can change the genetic constitution of the population. Complex systems, like those studied by biology, are going to harbor multiple, interacting forces at different scales, with variable temporal orders operating in diverse combinations in different particular situations. Integration of theories and models in such cases will not be as simple or global as in the case of, say, vector addition of electromagnetic and gravitational forces in physics.

Often, abstract modeling deliberately isolates the operation of a single causal process. How does mutation affect gene-pool frequencies, ceteris paribus? Or if not natural selection nor migration nor mutation is operating on a population, what will be the pattern of gene frequency changes over several generations? Theories abstract and simplify to achieve precision and generality. However, the concrete explanatory situations on which we bring the abstract models to bear are messy, perhaps unique products of historical contingencies and interacting, multiple causal factors. Of course, it is possible that a single causal process may completely determine a particular event. However, given the complexity and diversity of biological phenomena, simple models are more likely to capture partial causes. In this situation, individually more complete models may be developed if there are means for replacing the false assumptions in the model by more realistic descriptions (cf. Wimsatt 1987). More adequate description will require integrating diverse theories of all of the relevant features that uniquely characterize a given phenomenon, say, the ecological conditions of a particular lake. Hence, integration of diverse approaches is likely to be one hallmark of better science.

The work of Michael Friedman (1974) and Philip Kitcher (1981) seems to acknowledge the persistence of multiple potential descriptions of individual phenomena – the pluralism side of the story – but they marry it to a unificationist goal rather than an integrationist goal. Indeed, they argue that the unification of ever more phenomena under one theoretical schema is central to scientific explanation. This can be seen in Kitcher's criteria for identifying scientific progress with the acceptance of theories embodying the fewest explanatory schemata for the widest phenomenal coverage. This type of explanatory unification is certainly evident in some of the best-known cases in the history of science, including, famously, Newton's "unification" of terrestrial and celestial mechanics, Lyell's "unification" of the historically distant and current geological forces of change, and Darwin's "unification" of explanations of diversity and distribution of phenotypic features of humans and nonhumans by appeal to a single schematic principle of natural selection on heritable variation. Subsumption of diverse phenomena by appeal to increased generality and abstraction is indeed one way in which a plurality of accounts can be related. This constitutes a type of theoretical integration (see also Morrison 2000). However, I do not think it is the only mode of relationship. Therefore, identifying theoretical unification as the *only* means of doing good science is a mistake that removes the impetus to understand the value of diverse integrative strategies in scientific inquiry.

Darwin's great unifying insight in *The Origin of Species* (1859) was to see that the very different situations of the diversity of morphology of the thirteen species of finches he observed in the Galapagos Islands, the variety of domesticated pigeons in England, and the precise construction of the orchids he investigated were all explained by a single schema: natural selection operating gradually over slight heritable variation in individual members of a population.

While selection characterized at this level of abstraction unifies quite a significant number of phenomena, there are reasons to go both more abstract and less abstract with corresponding increases and decreases in generality. Within biology, hierarchical selection theory expands the biological targets from Darwin's individual organisms in a population both up to kin groups, trait groups, unrelated groups, and possibly species levels and down to gametes and genes. Darden and Cain (1989) move up the scale and beyond biology. "Natural selection, clonal selection for antibody production, and selective theories of higher brain function are examples of selection type theories. Selection theories solve adaptation problems by specifying a process through which one thing comes to be adapted to another thing" (106). (See also Skipper 1999.)

Within biology, even individual or organismic selection is broken down into r and k selection. R selection characterizes the processes occurring in populations that are not near the carrying capacity of their environment, and hence selection occurs by increasing fecundity, while k selection operates in populations at their quantitative limit, and hence the mechanism is instantiated by increased survival of offspring rather than number of offspring. The consequences of these different kinds of organismic selection are not always in concert. A number of studies both in the lab and in the field show mixed results of the relationship between fecundity and survival: out of 22 lab studies, a negative relationship was found in seven (and positive in five); out of 41 field studies, a negative relationship was found in 23 (and positive in four); the remaining cases were nonsignificant (Stearns 1992). Indeed, the more concretely described individual processes that constitute organismic selection may operate antagonistically, additively, or synergistically. Hence, collapsing the different processes constituent of natural selection operating on organismic variation in a population into a single representation of the process of selection can obscure what may be important differences. The point is that different levels of abstraction are required for different tasks (see also Mitchell 2000 for a version of this argument applied to chemistry).

Specific theoretical unifications, while being one form of integrating diverse causal models, must be justified by appeal to more than the fewer-the-better argument. That is, one must at least specify *what* it is "better" for.

Integrative Pluralism

Integration, the alternative to both reduction and isolation, occurs at many levels of abstraction and is driven by a variety of pragmatic interests. Establishing the philosophical arguments for the need for *some form* of integrative pluralism is clearly only the first step toward a better understanding of science. Indeed, the arguments I have given for expecting pluralism imply that the types of integration within science will also be varied and diverse. No single theoretical framework, no simple algorithm, will suffice. This is also evident from the type of case study–driven work that has already considered questions of integration. Darden and Maull (1977) outlined several kinds of "interfield" relations that might characterize integration across two theoretical boundaries: physical localization and part-whole relation, physical description of an entity described in another theory, identifying a structure underlying a function, and identifying a cause in one field of an effect described in another (see also Darden 1986, 1991). Bechtel (1986) identified other "cross-disciplinary" patterns that are nonreductive: conceptual links between disciplines to induce modification of perspective in one or the other, recognition of new levels of organization to solve unsolved problems in existing fields, using research techniques from one field to develop theoretical models in another, extending and applying a theoretical framework from one field to another, and "developing a new theoretical framework that will reconceptualize research in now separate domains as it tries to integrate them" (47). Bechtel and Richardson (1993) have further explored integrative strategies in their more recent work, especially with respect to crossing different levels of organization. These studies call for further cases in order to catalog what appears to be a wide variety of ways in which integration proceeds.

My own investigations of interdisciplinary work between developmental and evolutionary models (Mitchell 1992) and also between biology, writ large, and cultural models (Mitchell et al. 1997) have suggested a set of working hypotheses. In that work I proposed three types of integration: (1) mechanical rules, (2) local theoretical unification, and (3) explanatory, concrete integration.

Mechanical rules can be used to quantitatively determine the joint effects of independent additive causal processes explained by different theories.

Vector addition on the contributions of electromagnetic and gravitational forces to resultant motion is an example. The integration of theories is simply a demonstration that they are simultaneously applicable in a linear way. Sewell-Wright attempted to do the same for the effects of mutation and selection on gene frequencies (see Sober 1987). Prima facie, this type of integration seems appropriate for causes that are additive and operate on the same entities for comparable time periods. However, some biological phenomena do not seem to be amenable to mechanical rule integration. Think of the slime mold. When there is sufficient food in the environment, slime mold exists as independent amoebae. They move, feed on bacteria, and reproduce by cell division. When food becomes scarce, a chemical signal stops cell division and they move toward each other to form a multicellular "pseudoplasmodium" of tens of thousands of cells. This new association of cells then differentiates into a stalk supporting a fruiting body that produces spores. The spores are launched to spread to a new environment (with better chances of food), and life as individual amoebae begins again (Keller and Segel 1970). This type of "emergent" effect of the interaction of individual components of a complex system is typically nonlinear, and thus the individual component causes are nonadditive. Therefore, theories explaining the causes are *not* applicable without consideration of interactions.

The second model of integration, local theoretical unification, aims to develop models in which a number of features of a complex process are jointly modeled. This is similar to the explanatory unification counseled by Friedman and Kitcher. However, as I argued above, the appropriate scope of the unity and corresponding degree of abstraction will be settled by a combination of pragmatic and ontological constraints. The problem of scale in ecology illustrates this. In discussing the trade-off of detail for generality in modeling evolution on different classes of entity and ecological relationships among populations, Levin comments, "Here . . . the problem is not to choose the correct scale of description, but rather to recognize that change is taking place on many scales at the same time, and that it is the interaction among phenomena on different scales that must occupy our attention" (1992: 1947). Furthermore, "In general, one must recognize that different processes are likely to be important on different scales, and find ways to achieve their integration" (1950). A more concrete example of this is found in the discussion of "top-down" and "bottom-up" theories of the regulation of trophic structure and species composition in an ecosystem. Liebold et. al. (1997) argue that these "two artificially distinct perspectives" (468) that model the effects on trophic structure of predation and resources, respectively, are both supported by empirical evidence. They propose a "synthesized" model that takes both forces into account and thereby links community and ecosystem approaches. At the same time, they recognize that their model may still represent only "one subsystem in the more complex array of food web interactions," a local unification in my terminology, or, alternatively, may predict "the cumulative behavior of many subsystems that act in parallel but roughly additive ways" (483), thus constituting something like a mechanical rule.

The third type of integration – explanatory, concrete integration – appears to occur in cases of high complexity and pressing pragmatic goals (see Oreskes et al. 1994). That is, when a large number of at most partially in-dependent factors participate in structuring a biological process, and where those factors span time and dimension scales as well as standard scientific disciplines, even modest theoretical unification will be elusive. Think of the changes of state of a complex ecosystem such as that of Lake Erie. There is an ongoing modeling project to consider the lake-wide effects of the invasion of zebra mussels, declining phosphorus loading, continuing toxic contamina-tion, and fish harvesting on the structure of the fish community of that lake (see Culver 1999; Koonce and Locci 1999). The different factors contributing to these effects are large and diverse, including the chemicals silica, ammonia, nitrate, phosphate, total phosphorus, and phosphorus in sediments; five taxa of phytoplankton, six taxa of herbivorous zooplankton (including zebra mussel veligers), and three taxa of predatory crustacean zooplankton; zebra mussels and four taxa of other macrobenthos; and eleven taxa of planktivorous fish and six taxa of piscivorous fish. The model also represents seasonal and spa-tial variation in solar radiation. "Because phytoplankton, zooplankton, and planktivorous fish are sensitive to light, temperature, and nutrient concentra-tions they experience in non-linear ways the information represented in the model cannot be of 'average' values but rather of specific sampling locations in the lake" (Culver 1999). Features of the method of integration of these multiple factors for a single lake may be local to Lake Erie or may be symp-tomatic of a class of situations, but are unlikely to be global and algorith-mic. As such, I suggest, it may be better described as explanatory, concrete integration.

6.2. ON PLURALISM AND COMPETITION IN EVOLUTIONARY EXPLANATIONS

Introduction

The relationship between developmental and evolutionary biology is the focus of a recent pair of debates that surfaced in the journals *Natural History* and

194

Animal Behavior.[2] The first consisted of a heated exchange between Gould (1987a, b) and Alcock (1987) concerning the correct explanation for the clitoral site of female orgasm in humans and other primates. Gould identified this character as homologous to the male penis and appealed to "development" as the appropriate explanatory strategy with which to account for the female counterpart. Alcock suggested a possible adaptive function for clitoral orgasm, hence drawing on adaptationist models of explanation, that is, appealing to evolution by natural selection as the cause of the phenomenon.

The second debate, between Sherman (1988, 1989) and Jamieson (1989), was in response to the earlier exchange. Here the disagreement was about whether Gould and Alcock defended compatible or conflicting explanations. Sherman proposed a compatibilist reading of the issue, relying on a model of a plurality of levels of analysis. Jamieson disagreed.

Fundamentally, these controversies concern the possibility and mode of integration of developmental and evolutionary biology. In this section I argue that some of the confusion arising in the Gould-Alcock and Sherman-Jamieson debates can be eliminated by understanding scientific theories as idealizations. This view will allow us to see clearly when theories conflict and how differing theories may be integrated. To this end, I first present a brief account of the background and the content of the two debates and then turn to the question of how theories understood as abstractions or idealized models of causal relations may be integrated in the explanation of concrete phenomena.

Origin of the Problem

Gould's defense of a developmental explanation of clitoral orgasm is an instance of a general indictment of narrowing the range of acceptable explanations. In 1979, Gould and Lewontin's paper "The Spandrels of San Marco" urged a reevaluation of what they took to be the common explanatory strategy in evolutionary biology, namely, wanton adaptationism. They characterized this program in terms of two primary assumptions:

1. The evolution of complex organisms can be understood by the atomization of their phenotypic traits, and the explanation of each trait separately, and

---

[2] This section was originally published as S. D. Mitchell, "On Pluralism and Competition in Evolutionary Explanations," *American Zoologist* 32 (1992): 135–144. An early version of this section was presented at the Conference on Development and Evolution, organized by Stuart Kauffman at the Santa Fe Institute in October 1989. I thank the conference participants for their discussion. In addition I thank Nancy Cartwright, Michael Dietrich, Jerry Downhower, Philip Kitcher, Bill Wimsatt, and the anonymous referee for their comments on topics addressed in this section.

2. These atomized traits are explained by showing how they are globally optimal responses to selection pressures, that is, are adaptations in an engineering sense.

There are different interpretations of the message of Gould and Lewontin's paper. The most innocuous reading of this article shows Gould and Lewontin pleading for increased rigor in the testing and acceptance of adaptationist explanations, a message most evolutionary biologists find uncontroversial, if not unnecessary. Surely, Gould and Lewontin were suggesting something more. Another interpretation renders the paper as heralding an era of pluralism, where nonselective, or nondirectly selective explanations would be seriously entertained, accepted, and seen to complement adaptationist accounts. This is not a particularly new or radical claim, but is consistent with Gould's other attempts to "soften" the synthesis that had hardened in such a way as to ignore random and other nonselectionist sources of biological change. The strongest reading sees Gould and Lewontin claiming that nonadaptationist explanations should be preferred to adaptationist ones, and hence should replace them.

Clearly, the first interpretation suggests nothing substantial concerning the types of explanatory schema available or preferable for evolutionary biology. Rather, it repeats a general scientific indictment against hasty acceptance of any hypotheses without sufficient evidential support. It appears that this reading is at the back of Alcock's insistence that the adaptationist explanation for clitoral orgasm that he proposes be given the appropriate rigorous testing that any hypothesis requires (Alcock 1987).

The two remaining interpretations pose a greater challenge to adaptationist hegemony. If adaptationist explanations are to be either (1) accepted alongside alternatives or (2) superseded by them, we first must be clear about what constitutes alternatives to adaptationism. Two sources of nonadaptationist alternatives mentioned by Gould and Lewontin are chance and development.

Taking chance as providing a serious alternative to adaptationist explanations challenges common assumptions about the relative importance of genetic drift as a force in evolution. It is certainly possible that some form of sampling error (chance events skewing either the pool of adult reproductives or gametes) can intervene to change the terms under which nonchance events, such as selection, can operate. Given this, one is mistaken to accept without question that every sequence of evolutionary change, even in a fixed environment, is the inevitable result of increasing optimization. In *Wonderful Life* (1989), Gould emphasizes the role of chance extinctions, or historical contingencies, as a crucial component of our interpretation of fossils found in the Burgess shale, and hence of our understanding of specific evolutionary trajectories. By changing the landscapes on which selection may operate either by dramatically altering the environment or by randomly redefining the range of alternative phenotypes available to selection, chance clearly can play a significant role in long-term evolutionary processes. Although questions of degree may remain – specifying how frequently and how significantly chance events have influenced the course of evolution – the ways in which considerations of chance constitute real alternatives to adaptationism are clear. An explanation of a particular episode in terms of chance events excludes an explanation of that same event in terms of selection. (See Beatty 1987 for a clear analysis of how chance can be construed as an alternative to selection in terms of its prevalence in nature, its identification, and its pursuit methodologically for the community of biologists.)

In what follows I focus on the second source of nonselectionist explanations. In what ways does "development" contribute to or constitute an alternative explanatory program in biology? In addition to resolving the *Natural History* and *Animal Behavior* debates, clarifying this relationship is crucial to understanding more generally the integration of developmental biology and evolutionary biology (Maynard Smith et al. 1985). I take as my starting point a dispute that has recently concerned behavioral ecologists. One might subtitle this section, "How do adaptationists perceive the 'threat' from developmental biology?" Prior to deciding which of two explanations of a given biological phenomenon should be adopted by the scientific community, one must first decide whether the two are compatible or competitors.

Developmental Challenge

In his *Natural History* article "Freudian Slip" (1987), Gould argued that the clitoral site of female orgasm is a developmental artifact and not adaptively significant. Two months later, Alcock responded and Gould replied in the same journal. In 1988 this issue was placed in a more general context in an exchange between Sherman and Jamieson concerning "levels of analysis" that appeared in *Animal Behavior.* Though it is the latter discussion that raises the philosophical question of when two proposed explanations are in conflict, a brief description of the motivating problem will set the scene.

The message of Gould (1987a) is clear and familiar. Adaptationists wantonly invent selection scenarios to account for any apparent (to us) phenotypic trait. But clearly, this is not always justified. For example, it is accepted that male nipples are not adaptively significant for males because "males and females are not separate entities, shaped independently by natural selection. Rather the two sexes are variants upon a single ground plan, elaborated in late embryology" (Gould 1987a: 16). Selection on females accounts for the evolution of female nipples. Male nipples are functionless homologies maintained because they are a necessary consequence of the developmental sequence that results in female nipples. Similarly, Gould argues, female clitoral orgasm, which does not directly contribute to reproduction, plays no independently adaptive role for females. In this case, it is the male trait that is the appropriate candidate for a functional account, since it directly contributes to reproductive success. Female orgasm is the developmental, homologous artifact.

Gould documents that this explanation of clitoral orgasm escaped the notice of many biologists because they consider only adaptationist explanations as legitimate. Instead, they suggest a variety of fitness consequences that might plausibly account for the evolution of female clitoral orgasm, including, for example, its contribution to an increased emotional bond with a sexual partner. Clearly, from Gould's perspective, these proposals for the function of clitoral orgasm suffer from both adaptationist evils: misidentifying the correct target for natural selection and ignoring nonselectionist explanations. Thus, Gould (1987a: 21) says, "[M]any structures (such as male nipples and clitoral orgasm) have no direct adaptational 'why' . . . and we discover this by studying pathways of genetics and development . . . by first understanding how a structure is built."

Not all biologists were willing to accept Gould's pronouncements on the alleged homologues. Alcock (1987), while sympathetic to what he calls Gould's "plea for pluralism," rejects the substantial claim that an adaptive explanation of clitoral orgasm is unnecessary. He argues that the clitoris-penis relationship is not analogous to the male nipple–female nipple relationship because, unlike the female clitoris, male nipples are "inert" and do not "do anything." Female clitoral orgasm is not "inert." Neither should we think that it would have the same function as its structural counterpart in males. Given the differences in reproductive strategies of male and females, one should expect the functional roles of the organs and processes that contribute to sexual reproduction to differ correlatively. Hence, Alcock suggests that female orgasm may be used by the female to evaluate the probability of sequestering future parental care from a male. This presumes that the male's behavior involved in producing orgasm in the female is correlated with parenting behavior. Such a trait would then evolve via sexual selection. Alcock emphasized that this adaptation hypothesis, like any other, requires evidence from testing and should not be accepted by reason of mere plausibility.

Gould responded by suggesting that Alcock misunderstood the point. Even if it could be shown that female clitoral orgasm had such consequences in sexual selection currently, it may have no bearing on why such a feature had evolved historically. Organisms can make use of developmental by-products; hence, the current reproductive consequences do not show that a structure came to be present in a population as a result of its current function. Thus, Gould throws down the gauntlet once more: "Adaptations are features built by natural selection that enhance reproductive success. The domain of biologically useful structures is vastly greater.... developmental explanations are more expansive and operational than the necessarily fruitless and untestable adaptationist speculations" (1987b: 6). These are clearly fighting words, and Sherman (1988) stepped in to resolve the heated conflict by shifting the problem one level up. Must we decide between Gould's and Alcock's explanations of female clitoral orgasm? Are they, perhaps, both correct? More generally, we can ask, When are two proffered explanations incompatible alternatives, and when are they complementary components of a complete account of a biological phenomenon?

Levels of Analysis

The Local Argument. Sherman's paper "The Levels of Analysis" (1988) defends a pluralistic account of biological explanations that is designed to show how both Gould's and Alcock's explanations can be accepted. He suggests that the problem with the particular dispute over the evolution of female clitoral orgasms may be "semantic." "Adaptation" has a variety of meanings, and if two parties use the term in different ways in attempts to answer different questions, apparent disputes can be dissolved. Sherman is certainly correct that adaptive significance sometimes refers to the current effects on reproductive success of a trait and other times implies an etiology. (See Burian 1983 for a more detailed account of the various uses.) This latter view ascribes the term "adaptation" only to those traits that have evolved by means of natural selection. A trait is an adaptation only if it has become fixed within a population (or fixed at a specified frequency within the population) as a result of the greater reproductive success it conferred on individuals who possessed it in the past, relative to the individuals possessing alternative traits. That difference in reproductive success must be the result of interactions directly involving the competing traits in a shared environment. Sherman is also correct in claiming that Gould endorses the historical notion of adaptation (Gould and Vrba 1982). However, Sherman's attempt to defuse this particular debate by attributing the current function view of adaptation to Alcock fails. Alcock is explicit in adopting the historical account of adaptation. Current studies are conducted in order to acquire evidence for justifying such historical claims

(Thornhill and Alcock 1983: 2ff.). So Gould and Alcock really are debating what type of causal history is responsible for clitoral orgasm. Gould argues that only by understanding developmental structural constraints can one hope to propose a realistic hypothesis, and by so doing clitoral orgasm is explained by two processes: direct selection for the male trait and the developmental processes that issue in the female trait as a necessary consequence. Alcock suggests explaining the processes responsible for clitoral orgasm in terms of direct selection on females. They do not disagree on the question, but rather on what will constitute evidence for the various competing causal answers. Jamieson (1989) also chides Sherman for his misinterpretation of the Gould-Alcock debate as being a matter of mere "semantics." Though I clearly agree with Jamieson's conclusion, he bases it on a mistaken inference. He claims Gould is committed to viewing clitoral orgasm as currently selectively neutral and that Alcock views it as currently having important consequences for reproductive success. The historical dimension is absent from Jamieson's interpretation. In contrast, I have argued that Gould and Alcock are both attempting to explain past causal processes in terms of either past direct or indirect selection. Gould explicitly denies the claim that current functional consequences strongly imply past causal processes. He does not deny that current functional consequences might indeed be found.

The Global Argument. Although misconstruing the particulars of the Gould-Alcock debate and hence relegating it to a "semantic" dispute, Sherman's pluralist, compatibilist model of biological explanation must still be considered. Although it did not help in this case, it may sort real disputes between evolutionary and developmental perspectives from misdirected ones. "Clitorises, like other aspects of the phenotype, can be explained from several different, but not mutually exclusive perspectives" (Sherman 1988: 616). Thus, before biologists struggle over who has got *the* correct account, one must first ask whether advocates of developmental explanations and adaptationists are confusing different levels of analysis.

The hierarchical picture of biological explanations Sherman proposes is an elaboration of those suggested by Mayr (1961) and Tinbergen (1963) (see Table 6.1). Different levels represent divergent questions that one may raise. When the questions differ, then it is not surprising that the answers differ accordingly. Since the answers are not directed at the same target, they are not competing alternatives but, rather, compatible components of a multidimensional body of biological knowledge. Sherman revives Mayr's distinction between proximate, or "how," questions and ultimate, or "why," questions, accepting Tinbergen's elaboration of the original two categories into four. The four levels are: evolutionary origins, functional consequences

Table 6.1. *Sherman's Levels of Analysis*

Mayr	Tinbergen/Sherman	Question
Ultimate "why"	Evolutionary origin	Why did trait t arise in the lineage when it did?
	Functional consequences	What are the consequences of possessing t on fitness?
Proximate "how"	Ontogenetic processes	Why does t appear in the life-cycle when it does?
	Mechanisms physiological	What hormonal mechanisms trigger expression of t?
	cognitive	What mental functions control expression of t?

Note: This table expands Sherman's (1988) table by adding the question column. In addition to a cognitive component at the level of mechanisms, one might also include cognitive development under ontogenetic processes.

(the two "why" questions), ontogenetic processes, and mechanisms (the two "how" questions). Sherman updates Tinbergen's categories, bifurcating questions of mechanisms into those that target physiological processes and those that target cognitive processes. With respect to this classificatory scheme, Sherman claims, "Every hypothesis in biology is subsumed within this framework; competition between alternatives appropriately occurs within and not among levels" (Sherman 1988: 616).

The strategy of defusing alleged disputes by clarifying the questions that the disputants are attempting to answer has much appeal (see Mitchell 1987b). The pragmatics of explanation (Belnap and Steel 1976; Garfinkel 1981; Van Fraassen 1980) provide a framework in which to show how similar linguistic locutions sometimes hide important differences in explanatory projects. For example, one may ask, "Why did the Cape May warbler fly north on May 2, 1991?"

The context in which the question is raised is required in order to distinguish between two separate explanatory projects that might be expressed. In the above example, one may be inquiring into (1) the conditions that triggered the migration on that precise date, or one may be wondering why (2) the Cape May warbler migrates at all. The implicit parts of a question, usually present in the context in which the question is raised, can be made explicit by specifying the intended contrast class and the relevance relation that is to hold between the answer and the question. The contrast class lists the events excluded by the explanation, that is, what would follow an explicitly stated "rather than." For (1) it would be "migrating on April 30, on May 1, on May 3, on May 4, etc." For (2) it would be "nesting in the present location, as do black and white warblers. " What makes an answer relevant to the intended question is also part of the unexpressed context. That relation may be the proximate triggering cause of a singular event or the historically plausible conditions causing a property to be typical of a population.

By making explicit the implicit context of a question, heretofore hidden differences can be usefully exposed. If Gould and Alcock used "adaptation" differently, then Sherman's account would have done just that. Contrary to Sherman, I have argued that in the case of the Gould-Alcock debate, just such a clarification led to the conclusion that, in fact, the two disputants were addressing the same question, with the same contrast class, namely, the presence, rather than the absence, of female clitoral orgasm in populations of primates. They seem also to agree that the relationship that constitutes a relevant answer to that question is the causal process or processes that have been responsible in the past for bringing about this phenomenon. They disagree not with the question, but rather with the answer to that question. If a biological phenomenon is completely explained by appeal to natural selection operating on it directly, as Alcock has suggested in this case, then it could not at the same time be truly explained by indirect selection, or a nonselective process, which is Gould's view. What is required in order to sort debates into the "merely semantic" and the substantial is a classification of answers, as well as a classification of questions.

One way to identify substantial differences is to characterize the answers as appeals to distinct abstract models of causal processes and the specifications for application to a given case. I endorse the view of Cartwright (1980) that the scientific laws used in explanations of concrete phenomena must be interpreted as having a ceteris paribus clause. Namely, they describe what is to be expected in idealized situations, when only one of a set of potential causal factors is operating, that is, when nothing else is interfering. Given that we accept that multiple causal factors can and often do interact in the production and maintenance of a given trait in a population, without such a proviso, the laws would be literally false. With the proviso, however, such models do not themselves directly account for many, if any, real cases. This is because the concrete explanatory situations on which we bring the abstract models to bear are messy, perhaps unique products of historical contingencies and interacting, multiple causal factors. Wimsatt (1987) develops a similar point with respect to the use of false models in science. Clearly, not all false claims are useful. Nevertheless, simplified, incomplete, or idealized models that characterize a complex situation in terms of its component parts may lead to more complete and realistic explanations. For this, one needs methods for developing and applying models that replace the false assumptions by more accurate descriptions. (See also Richerson and Boyd 1987.)

It is my thesis that to locate the domain of disagreement between so-called adaptationist and developmental explanatory strategies, we need to distinguish the idealized models from the particular application of such models to a concrete case.

Idealized Causal Processes

Any single causal process may completely determine or partially contribute to trait t's presence in population p. Abstract models describe the effects of the operation of a specific causal process. Candidates for component causes in the evolution of a complex trait are as follows.

Direct Selection for the Adaptedness of the Trait. Here selection operates on variations in the trait and, when such variation is genetically transmitted, will issue in differential representation of the trait that is more adaptive, that is, whose consequences in interaction with the environment enhance the reproductive success of the trait bearers. Consequences on reproductive success are thus causally significant, and the trait is an adaptation. However, it is the past consequences that determine present features, and these are not necessarily identical to the current observed consequences. This is the model behind Alcock's hypothesized answer to the question of the evolution of clitoral orgasm.

Indirect Selection of Part of an Adaptive Complex. In this case a set of traits is seen as homologous in the sense that they are produced by one and the same developmental process. Two or more traits are welded together by means of molecular relationships (e.g., genetic linkage) or other developmental processes (e.g., pleiotropy). In our example, this would mean that you don't get female nipples without also producing male nipples. But a dependent part of the complex, male nipples, may not be the object of selection directly. Rather, some other aspect of the complex, the expression of nipples in females, may be selected. Because of the way in which molecular and developmental processes connect these conceptually distinct aspects of phenotypic expression, the entire complex will replicate as a result. Hence, some component of the suite, expressed in the phenotype, may evolve as the indirect consequence of selection on a different part of the suite. This appears to be Gould's hypothesis to explain the evolution of clitoral orgasm. Until we know what the developmental processes are, we will not be in a position to accurately identify the results of selection on integrated complex phenotypes.

This account of the developmental alternative to adaptationism gives development the role of constraining the objects and consequences of selection operating on a phenotype. Identifying the developmental relationships is not contrary to a selectionist account, although it may challenge a claim about the objects toward which selectionist explanations are directed.

Nonselective Explanations. (1) Chance, (2) migration, and (3) mutation can all be processes that issue in the change of frequency of a trait in a population, that is, its evolution. There are conditions, such as size of the population, that make the operation of these processes more or less probable and hence the applicability of such answers to the evolutionary question more or less reasonable.

(4) Direct environment modulation. The presence of phenotypic variation within a population at times may not reflect variation at the genetic level. This can occur when direct changes in local historical experience influence the variants. Such changes will not be preserved over time, but if the environmental conditions that directly induce the phenotypic properties continue to occur, it may appear that the variation is transmitted rather than reinvented in each generation. Cultural selection may be one source of direct environmental modulation. Jamieson (1986) outlines such environmental and developmental alternatives to selectionist, or what he calls "functionalist," explanations of behavioral phenotypes. "New behaviors may persist in a population either because of changes in the average genotype by natural selection or by enduring changes in the environment in which the average genotype develops" (Jamieson 1986: 98; see also Koenig and Mumme 1990).

(5) Developmental laws of form. While an understanding of *bauplane,* or developmentally connected suites of traits, may assist in identifying the correct target of selection, as such it does not present an independent causal process for generating change in a population's trait frequencies. This is the case of indirect selection discussed above, where natural selection remains the motive force.

However, some have suggested that developmental causes may be responsible for the presence of traits *despite* the operation of selection. For example, Kauffman (1987, 1993) has developed abstract Boolean connected-network and dynamical models to show what kinds of order emerge "spontaneously" in complex systems. His theory presents an alternative causal model to selection for complex systems by showing what specific traits may be self-organized consequences emerging from the basic structure of the system itself and not the direct or indirect product of natural selection. In fact, if the kinds of features identified by Kauffman as "typical" of classes of systems are fairly ubiquitous, the variation in traits required for the operation of selection would be either unlikely or completely absent.

Page and Mitchell (1991) have applied this type of model to colony-level phenotypic traits of honeybees. By so doing, they have argued that division of labor and task specialization, traits traditionally thought to be the result of evolution by colony-level selection, may instead be generated automatically with the emergence of complexity in the social insects. By presuming that individual bees have the properties known to be possessed by solitary bees, that there is genetic variation between individuals, and that there is minimal toleration of the presence of other individuals, Page and Mitchell show how division of labor and task specialization arise "spontaneously." The energy efficiency of colonies with these features compared with those without, the basis for conjectured selection etiologies, may not explain the presence of the traits, if such variation is unlikely given minimal sociality.

Mathematical elegance and tractability, computer simulation, and laboratory experimentation will all contribute to the development and specification of the individual causal models. However, it is unlikely that the ideal circumstances modeled separately have occurred in such ideal form in nature. Nevertheless, any single model may capture some part of the actual causal history of a specific phenomenon. For example, it may be the case that direct selection has worked in conjunction with drift, or that crude division of labor arose spontaneously in honeybees but was honed and developed by colony-level selection on more limited variations.

Pluralism or Competition?

Sherman's appeal to levels of analysis argues for a plurality of legitimate questions and a limitation on the conflicts that can occur among biological explanations to those occurring within a given level. While some biologists may restrict their interests to a specific level, this is not necessarily the case. There are genuine disputes about biological explanations, and the recognition of disputes and attempts to resolve them affect both scientific theories and experimental practice. As we have seen in the Gould-Alcock debate, one may be concerned not exclusively with the narrow question of, say, a trait's current function, but rather with how those effects may enter into the complex process that explains or fails to explain the evolutionary etiology of the trait.

The attraction of pluralism is not, pace Sherman, its ability to represent compatible answers to disparate questions. Pluralism is better applied to the causal models that by modeling the contribution of specific causes abstract away the operation of other compounding factors, and hence make no incompatible claim about their operation. Once the structure and application of causal models is recognized, one may see that while disputes may indeed arise within a level with respect to the details of a given model (like how best to measure fitness in determining the current function of a trait), answers at different levels are not unrelated. Thus, although pluralism within biology is to be applauded, it is not the pluralism of questions and the consequent independence of answers that Sherman endorses, but rather a pluralism of models of causal processes that may describe contributing factors in a given explanatory situation. To argue for the latter type of pluralism, while recognizing competition in the explanations of particular phenomena, requires a further account of how idealized models are to be integrated in concrete, nonideal cases.

There are three candidates for integrating theories that might apply to the cases described above, that is, that of development and selection. There are (1) combinatorial rules, (2) unified theory, and (3) piecemeal integration. I discuss the applicability of each in turn.

Combinatorial Rules. Vector addition in physics provides a general method for combining the effects of independent forces of, say, gravity and electromagnetism on the motion of a body. Sober (1987) has shown that this type of integration may be adequate to predicting and explaining the joint effects of mutation and selection on gene frequencies. However, the effects of developmental *bauplane* and selection are not additive. Sober (1987) considered and rejected such a mechanism for similarly combining the effects of pleiotropy and selection. A common currency in which to measure the effects of contributing causal processes is necessary in order to use a combinatorial algorithm. Gould's and Lewontin's arguments for recognizing development in evolutionary explanations suggest that one role that developmental processes may play is in actually changing the objects on which selection operates. For this reason, it seems that this mode of integration will not be universally applicable.

Unified Theory. Some have counseled the strategy of progressively increasing the realism of the assumptions made in the abstract theories that model specific causal processes (Burian 1986; Richerson and Boyd 1987). The aim of this type of integration appears to be the development of a grand unified theory of biology in which all aspects of the complex processes are represented. Various forms of reductionism can be understood as attempts at unification. For example, while one may model the effects of genetic drift when no selection is operating, this would be seen as a temporary step toward formulating a law in which all potential causal factors would be represented. There is no prima facie reason to reject this aim. Indeed, theoretical consistency among the various causal models is certainly a scientific virtue. However, to date no grand theoretical unification has developed. There may be some reason to believe that a single formulation will not be found to adequately represent all contributing causal factors. There do not appear to be univocal rules governing the nature of developmental constraints upon selection (see Raff et al. 1991). While such constraints may at times limit the power of selection to change gene frequencies, by proscribing some combinations as unavailable alternatives, at other times selection may act to change those very constraints. The conclusion to be drawn is that, at least at present, there is no strong indication that a unified theory is in the wings.

Piecemeal Integration. Without combinatorial algorithms or a unified theoretical framework for integrating the multitude of causal processes contributing to complex biological phenomena, one is left with a piecemeal approach to integration. This view recognizes the plurality of models (contrary to the unification goal), the nonindependence of at least some of the contributions to complex combinations (contrary to the algorithm goal), and at the same time the singularity of the particular combination of causal processes responsible for any given phenomenon.

The absence of either of the generalized modes of integration does not prevent the sorting out of debates such as that of Gould and Alcock concerning the evolution of clitoral orgasm in female primates. Using a piecemeal approach, one can acknowledge that the general explanatory strategies appealing to specific causal processes make simplifying assumptions that are literally false in describing an actual causal history. Nevertheless, the ways in which the model is false in a particular case may be conjectured and tested so that a more realistic description can be generated for the specific case. One expects neither that the actual processes in need of explanation will directly reflect the content of the models nor that the theoretical models will be modified to incorporate the details of all potential causal factors in a general framework. Rather, the actual configuration of complex processes resulting from development, selection, environmental modulation, and chance must be determined on a case-by-case basis. Thus, pluralism at the theoretical level does not entail pluralism with respect to the explanations of specific cases. There is indeed a conflict between Gould and Alcock. While each accepts the potential relevance of a plurality of causal processes, including adaptation and development, they disagree about which particular combination of forces is responsible for the causal history of clitoral orgasm.

207

6.3. INTEGRATIVE PLURALISM

The "fact" of pluralism in science is no surprise.[3] In scanning contemporary journals, books, and conference topics in some sciences, one is struck by the multiplicity of models, theoretical approaches, and explanations. Yet if science is representing and explaining the structure of the *one* world, why is there such a diversity of representations and explanations in some domains? One response is that pluralism simply reflects the immaturity of the science (Kuhn 1962). Yet history shows us that many sciences never exhibit a diminution in the multiplicity of theories, models, and explanations they generate. This "fact" of pluralism, on the face of it, seems to be correlated not with maturity of the discipline but with the complexity of the subject matter. Thus, the diversity of views found in contemporary science is not an embarrassment or a sign of failure, but rather the product of scientists doing what they must do to produce effective science. Pluralism reflects complexity. What type of pluralism? What type of complexity?

In this section I consider several philosophical explanations of scientific pluralism. I then defend an integrative model for understanding compatible alternatives and illustrate my thesis by analysis of explanations of division of labor in social insects.

Competitive versus Compatible Alternatives

To understand scientific pluralism, we need the distinction between what I call competitive and compatible pluralism (Mitchell 1992; Mitchell et al. 1997). Almost all recent philosophers of science concerned with pluralism have concentrated exclusively on multiple, competing hypotheses, such as the wave and particle theories of light or Darwinian and Lamarckian theories of inheritance (for exceptions, see Cartwright 1994; Dupré 1993, 1996; Hacking 1996). Feyerabend (1981) and Lakatos (1978), writing in the Popperian tradition, endorsed the desirability of maintaining a number of competing research programs or theories in order to hasten progress in science. The argument is that competition among alternative accounts of the same set of phenomena presents the most severe environment for the testing of any individual account. For these writers, scientific growth occurs by exposing false hypotheses

[3] This section was originally published as S. D. Mitchell, "Integrative Pluralism," *Biology and Philosophy* 17 (2002): 55–70. Research for this section was supported by the Wissenschaftskolleg zu Berlin and by the National Science Foundation, Science and Technology Studies Program, grant no. 9710615.

or weak portions of a research program to empirical scrutiny and then rejecting the offending hypothesis or strengthening the weak assumption. In this
framework, having a plurality of competing alternatives is supposed to increase the probability that particularly troublesome empirical results for any
given hypothesis will be exposed.

Competitive pluralism also has been defended from a different perspective, by Beatty (1987) and Kitcher (1990). For them, maintaining multiple,
competing theories and explanations is deemed the rational strategy to adopt
for the scientific community as a whole in order to hedge its bets against
empirical uncertainty. Given our epistemological fallibility, one cannot reliably project that the theory that accrues the highest epistemic warrant based
on current empirical support will continue to win out in the light of future
findings.

So, while these two approaches tolerate competing views, science nevertheless is thought to aim at ultimately resolving the conflicts by adopting the
unfalsified, or "true," or overwhelmingly supported winner of the competition.
As Kitcher puts it, "[t]he community goal is to arrive at universal acceptance
of the true theory" (1991: 19). The slogan here might be "Pluralism: The Way
to Unity." These accounts of competitive pluralism lead to the judgment that
pluralism is temporary and strategic, but ultimately eliminable. While this
correctly identifies some of the diversity of models and explanations found in
contemporary science, it fails to capture all of it. The remainder is constituted
by compatible, not mutually exclusive, alternatives.

Biologists have long recognized the diversity of compatible explanations.
Mayr (1961, 1982) distinguished between *how* and *why* questions that one
may ask of nature. This distinction was introduced in an effort to protect the
autonomy of separate biological traditions from a perceived threat of reduction and elimination (Beatty 1994). Later, Tinbergen (1963) outlined a four-
part classification of questions one might ask of a biological phenomenon.
Recently, Sherman (1988; see also Reeve and Sherman 1993) revived these
approaches under the name "levels of analysis." Biological questions are sep-
arated into levels of evolutionary origin, current reproductive function, ontogeny, and mechanism. Questions at different levels require different answers
(see Van Fraassen 1980). Answers could only compete, the argument goes, if
they were addressed to the same question or level of analysis. Sherman says,
"Every hypothesis within biology is subsumed within this framework; competition between alternatives appropriately occurs within and not among levels"
(1988: 616). While there are valuable insights in this account of compatible
pluralism, *the levels-of-analysis framework fails to adequately represent the
relations between alternative explanations.* I argue that it misconstrues where

conflict does and should occur and where alternatives are correctly judged to be compatible. In the extreme case, it can lead to a form of isolationism that can impede answering questions within any single level. The mistake lies not in recognizing a diversity of questions – indeed, scientists do pose a variety of questions to the subjects they study – but rather in the assumptions made about the epistemological structure of the answers. In other words, even when the questions that scientists pose are disambiguated, there remains a variety of compatible answers.

The Case of Division of Labor

To show that the levels-of-analysis model of compatible pluralism is mistaken and to defend an alternative view, I consider the example of explanations of division of labor in social insects. A social insect colony is a complex system, in several ways. Compositionally, it is complex, being constituted by a whole made up of many, nonrandomly structured parts. Dynamically, it is complex, in that it is the location for multiple, interacting causes, some of which are represented as linear, some as nonlinear functions. And evolutionarily, it is complex, in that social insects display a variety of historically contingent, adaptive responses to environmental challenges. Given the irreversible nature of the processes of evolution, the randomness with which mutations arise relative to those processes, and the modularity by which complex organisms are built from simpler ones, there exists in nature a multitude of ways to "solve" the problems of survival and reproduction. Relative, not optimal, adaptive peaks characterize the direction in which evolutionary change occurs. This implies that diverse adaptations, even under the same or similar environmental conditions, are likely to be found in nature.

A social insect colony consists of tens of thousands of individuals engaged in a variety of behaviors, including cell cleaning and capping, brood and queen tending, comb building, cleaning, food handling, guarding, and foraging (see Winston 1987). The individuals are not randomly engaged in tasks, nor does each individual do all the jobs available. Rather, there is a pattern of distribution of behaviors called "age polyethism" that changes through the life of the individual: young individuals work inside the hive, older individuals work outside. For bees, four age "castes" have been distinguished in terms of behavioral variation: (1) cell cleaning, (2) brood and queen care, (3) food storage, and (4) foraging. Each "caste" consists of a repertoire of tasks, and the individuals vary in their degree of specialization within a caste set. In addition, the colony as a whole displays plasticity by adjusting the proportions of workers active in particular tasks in correspondence to both internal and external factors. If there is a loss of foragers, the younger individuals may leave their nursing or food-storing tasks to fill the vacant jobs. This homeostatic response is accomplished in different ways by different species. Honeybee colonies harbor sufficient genetic variability among the workers to generate variant responses to stimuli (Calderone and Page 1992; Page and Metcalf 1982). Ant colonies, on the other hand, are generally not genetically diverse (but see Boomsma et al. 1999 for the exception in ants) and thus may accomplish the same sort of response flexibility by means of nest architecture (Tofts and Franks 1992) or some other mechanism. The ways in which information gets modulated through these systems depends on a host of other properties. What is important to notice is that while one species may solve the same problem one way, another may solve it in a different way.

The standard account of the ordered complex pattern of division of labor has been an adaptationist one (Wilson 1971). That is, the patterning of work within a colony is analyzed to determine what would be optimal in terms of ergonomic efficiency. If other than the optimal patterns had been exhibited in the past, they would have lost out in the competitive struggle of natural selection operating at the colony level on heritable variation. Thus, the one we see – the optimal, age-related one with specialization – would become stabilized by natural selection sorting out inferior arrangements.

The adaptationist explanation of division of labor "black boxes" the mechanisms by which the pattern is generated. It ignores, if you like, the physiology or development of the colony phenotype. By so doing, it implicitly assumes that however a phenotypic trait may come to be expressed, as long as it is heritable, natural selection would be able to optimize on variations of that trait. However, how a trait develops can and does restrict the range of adaptive explanations that could be plausibly entertained. In particular, self-organization models of aggregate or emergent traits provide for the possibility that natural selection alone is not the appropriate agent to explain some complex traits. Three such self-organization explanations have been recently developed for accounting for division of labor. These all challenge a pure colony-level selection account, since they all suggest that colony-level traits such as division of labor and specialization may be, at least in part, automatic expressions of the interactions of actors at the lower level.

Genetic Diversity for Threshold Response. Page and Mitchell (1991, 1998) ran simulations of honeybee colonies inspired by Kauffman's $n\text{-}k$ Boolean network model of complex systems (Kauffman 1993). Each colony had n individuals that interacted with k others. The interactions were a subset of the Boolean functions that characterize threshold functions by which the input presented to an individual would generate a response either to forage or not to forage. The model presupposed genetic variation among the individuals that controlled the threshold level required for response to a fixed stimulus. When presented with a given level of stimulus, the bees would self-organize into a pattern of working and resting and by working (e.g., foraging) would change the stimulus level that would be presented to the next bee. What emerged within this simple model of amplifying feedback process was task variation within an age caste and the tuning of behavior to colony needs. Specialization occurred "spontaneously" from mere genetic variation. The conclusion drawn was that some aspects of division of labor can arise – indeed, must arise – among groups of cohabiting and mutually tolerant individuals who harbor genetic variation.

The assumptions of this model are

1. Individuals vary genetically in their threshold of response to stimuli.
2. The initial distribution of the threshold is random.
3. Behavior moderates subsequent stimuli.

A pattern of division of labor and specialization coordinated to colony needs emerges from self-organization.

Recently, Fewell and Page (1999) empirically tested this model by forcing a solitary ant species, *Pogonomyrmex barbatus*, to cohabitate. They compared the results on task specialization with *P. californicus*, a related, but naturally cooperative ant. They concluded, "[O]ur results provide strong evidence that task specialization is an emergent property of sociality" (543).

Foraging-for-Work Algorithm. Tofts and Franks (1992) proposed a "foraging for work" algorithm as the basis from which division of labor emerges (see also Bourke and Franks 1995). In this model all the individuals operate with an identical algorithm, namely, to actively seek work. If work is found in one step of a clocked interval, then the individual performs the task; if not found, then it will move to another work station. The individuals are all born into one station, and the work stations are organized linearly, akin to a mass-production line in a factory. After an individual has moved to the second station or beyond if no work is found, it is randomly assigned to move either upstream or downstream. What emerges from this simple set of rules is the correlation of work load to task need. An efficient distribution of workers to task is achieved, and epiphenomenally age distribution falls out.

The assumptions here are that (1) individuals have identical work algorithms, and (2) nest architecture (distance from brood pile) varies the stimulus presented. A pattern of division of labor coordinated to colony needs and age polyethism emerges.

212

Learning Algorithm. Deneubourg et al. (1987) offer yet another self-organization model by which to explain the division of labor. In this case, neither genetic variation nor nest architecture drives the pattern we see, but rather it is individual learning, which operates as an amplification mechanism. Here the model assigns an identical learning algorithm to all the individuals, but assumes asynchrony in the birth of new workers. There are n foragers in two zones. Each individual (ant, in this model) has an assigned probability of foraging at all and a specified probability of foraging in zone 1. At the outset, all the individuals have identical probability assignments. However, through learning and forgetting the probabilities are changed.

Thus, the "interplay between amplification mechanism, competition between different pieces of information and individual randomness are the factors that generate the individual behavioral patterns and the society organization" (1987: 249). What emerges again is a pattern of division of labor with age polyethism. Here the assumptions are: (1) individuals have identical learning algorithms, and (2) there is asynchrony in the birth of new workers. A pattern of division of labor with age polyethism emerges.

The three self-organization models provide a very small subset of the theories and explanations found in contemporary biology to explain division of labor in social insects. In addition, a variety of different factors and features are appealed to, including: physiological changes, such as glandular development and reabsorption, where changes in body mass are coincident with changes in task; quantity of juvenile hormone secreted associated with tasks; colony population size and character, influencing foraging age; quality of available resources of nectar and pollen, affecting patterns of division of labor; and natural selection, operating at the colony level for greater efficiency.

What the self-organization models suggest is that the phenomenon seen at the social or colony level is not necessarily determined by a genetic blueprint in the individuals, which could be optimized via colony-level selection. There is no necessity to postulate an internal clock that determines which individuals engage in which tasks at which times. Rather, the patterns can be generated by means of the interaction of individuals on assumption of a very few internal components: genetic diversity in threshold response, uniformity of work, or learning algorithms. With the development of theories of self-organization, the toolbox of explanatory models available is expanded. Now not only are there optimality models for adapted evolution at the various levels of organization: the "selfish gene," the cell lineage, the individual, the kin group, the colony, and the species. But in addition, the phenotypic characters we seek to understand may be explained by self-organization of components of complex systems.

Table 6.2. *Where Conflict Occurs*

Can colony-level natural selection (evolutionary level) conflict with self-organization model (ontogenetic-level) explanation of division of labor?	
Levels of analysis	No
Integrative pluralism	Yes
Can genetic, architecture, and learning models of self-organization (all ontogenetic-level) conflict in explaining division of labor?	
Levels of analysis	Yes
Integrative pluralism	No

The obvious questions arise. How are these different explanations and different models of self-organization related? Are they mutually exclusive competitors, are they equal components of a grand unified theory, or are they compatible alternatives in a pluralism of explanatory resources?

In Sherman's levels-of-analysis model, conflict between diverse explanations should occur only within a level and not between levels (see Table 6.2). Is this correct? Self-organization models of division of labor would be located at the level of ontogeny, colony-level selection models at the level of evolution. For Sherman, no conflict should occur between them. However, self-organization models claim that, given certain properties of the individual components, interactions in a complex system necessarily generate certain forms of order. If this is the case, then historically there would have been no variation among social insect colonies for division of labor per se, since any complex so structured would display this pattern. Yet the evolutionary explanation of the origin of division of labor appeals to colony-level selection for energy efficiency and thus must, by definition, presuppose a history of heritable variation between colonies with such a pattern and colonies without it. The answers at the two levels, ontogeny and evolution, are not mutually consistent. Thus, contrary to Sherman, there can be competition between levels. Solutions to questions at one level, the developmental, affect the set of possible solutions at another level, the evolutionary. There are two different ways in which this could occur. On the one hand, developmental explanations might limit the range of viable variations that can be postulated for natural selection to have sorted between. Thus, natural selection could fashion not any logically possible response to an environmental "problem," but only those responses that meet the developmental constraints. On the other hand, developmental theories might discover structural necessities or universals. This implies not just a limit on the range of variants under selection but an absence of variation itself. If so, then natural selection could not be the causal agent by which to explain the presence of the trait we now see. A classic example for this type of constraint is found in Murray's reaction-diffusion model explaining why a spotted animal can have a striped tail but a striped animal cannot have a spotted tail (see Murray 1988; Oster and Murray 1989).

The other implication of Sherman's framework is that competition among explanations occurs within a level. If that were the case, we would expect that the three self-organizational models, all located at the ontogenetic level, would be mutually exclusive alternatives. But are they? I think not. An explicit feature in each of the models is the focus on a single causal factor and the exclusion of other possibly contributing causes. The different models of self-organization are idealized abstractions from the actual, complex phenomenon of division of labor (see Fig. 6.1). That is, the genetic model of Page and Mitchell ignores all factors but individual genetic diversity for threshold response for foraging.

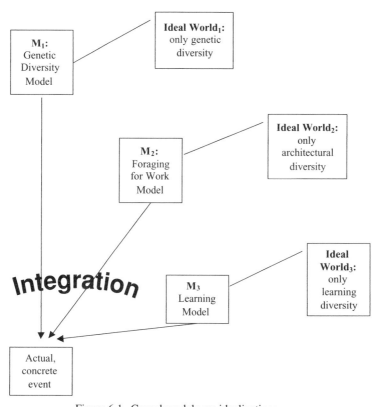

Figure 6.1. Causal models are idealizations.

In effect, it accurately describes only ideal cases where those simplifying assumptions hold true, but only partially captures actual cases that are not ideal in this way. So, too, for Tofts and Franks's foraging-for-work algorithm and Deneubourg et al.'s learning algorithm. As Deneubourg et al. state, "[t]he mathematical model presented takes only learning into account, and ignores all other factors, such as age or genetic differences, which could be involved in the foraging patterns. This attitude is justified on the one hand by lack of data and on the other hand in that we wish to understand the role and limits of learning as a mechanism contributing to an insect society's organization" (Deneubourg et al. 1987: 179). Contrary to the levels-of-analysis picture, the three self-organization hypotheses, although located at the same "level" of analysis, do not directly compete, since they describe only what would happen in nonoverlapping ideal worlds. How then can the diversity of explanations and models of division of labor be represented to better illustrate real conflict and the nature of compatibility?

Integrative Pluralism

Since the levels-of-analysis taxonomy in terms of different questions and correspondingly different answers failed to identify the location for competition, what then forces choices between different explanations? The answer, I believe, lies in the distinction between the theoretical modeling itself and the application of any model or models to the explanation of a particular, concrete event. As we saw above, at the theoretical level, pluralism is sanctioned. At the concrete explanatory level, on the other hand, integration is required. However complex and however many contributing causes participated, there is only one causal history that, in fact, has generated the phenomenon to be explained. Return to the case of division of labor. In honeybees, the queen mates with up to seventeen different drones, thereby producing genetic diversity among the workers in the colony (Page and Robinson 1991; Page et al. 1989b). The hive itself is concentrically structured, with the brood in the center, storage cells further out, and the entrance for foragers at the extreme edge. And the individuals are born at different times, giving them different learning opportunities. Thus what accounts for division of labor in the case of honeybees will be, presumably, a combination of genetic, learning, and architectural causal components. The three self-organization models include assumptions whose applicability to this case is testable and determinate. In addition, the selection may act on the features of this trait that remain variable and thus fine-tune features of division of labor to the selective environment. For example, while genetic variability will generate some pattern of specialization, selection can operate on the degree of variability as well as the components of variability if these turn out to influence colony fitness.

If different models are perceived as partial solutions to a biological question, then one might argue that a theory of division of labor would be one that correctly unified the partial accounts. However, while integration of the partial accounts is, indeed, required for explaining a concrete particular, unification at the theoretical level is unlikely to be very robust. This is due to the nature of the complexity characterizing the domain of phenomena studied by biology. It is the diversity of the "solutions" to adaptive problems and the historical contingencies influencing those variable paths that preclude global, theoretical unification. Ants, for example, exhibit division of labor similar to bees. Ant colonies, however, contain negligible genetic diversity compared with bee colonies. Thus, the theoretical constituents that would be integrated in the explanation of division of labor in ants would not be the same set as those required for the explanation of the "same" phenomenon in bees. For ants, the genetic self-organization model would often not apply. Even that claim is contingent, since Boomsma et al. (1999) has detected in one species of ant nearly as much genetic diversity as in honeybees. Thus, what in fact explains the existence and characteristics of division of labor in social insects, what appeared at first sight to be the "same" phenomenon requiring a single explanation, will be itself contingent on the particular features and pathways that occur in each case. At the same time, there is only one "true" integrated explanation for honeybees and one "true" integrated explanation for leaf-cutting ants, and hence competition among explanations in specific cases will and does occur. Nevertheless, compatible pluralism will remain for the models of potentially contributing causes, if not in the application of the models to explain a specific, concrete event.

Conclusion

Complexity in the sense of the diversity of the contingent, evolved properties of biological phenomena has important implications for how we understand the relationships among the plurality of theories and explanations found in contemporary biology. I have argued that there is no reason to view all forms of scientific pluralism as embarrassments or signs of immaturity. This latter perspective rests on mistakenly interpreting alternative theories and explanations as always competing. I have shown that, while some of the plurality found in biology is of this kind, many models used in explanation and prediction are not competing, but are compatible. In these cases, competition still occurs, but it is generated only in the integrative application of the models in explaining particular cases.

In defending integrative pluralism, an image of science that makes room for compatible pluralism, I have attempted to steer clear of two undesirable methodological pitfalls. The first is an isolationist stance that partitions scientific investigations into discrete levels of questions and their corresponding answers in a way that precludes the satisfactory investigation of any of the levels. The second is an uncritical anarchism that endorses all and any propositions. Neither of these positions correctly locates where and when competition in fact occurs between theories and explanations in biology. I have appealed to the idealized structure of scientific models and emphasized the distinction between the model and its application to a concrete situation. While the idealized and abstract character of models allows compatibility at the theoretical level, the realistic and concrete nature of explanation entails integration and resolution. Given the multiplicity of causal paths and historical contingency of biological phenomena, the type of integration that can occur in the application of models, that is, their use in explanations, will itself be piecemeal and local. The result is that pluralism with respect to models can and should coexist with integration in the generation of explanations of complex and varied biological phenomena.

References

Alcock, J. (1987). "Ardent Adaptationist." *Natural History* 96: 4.

Allen, G. (1978). *Life Science in the Twentieth Century*. Cambridge: Cambridge University Press.

Anderson, R. H. (1963). "The Laying Worker in the Cape Honeybee, *Apis mellifera capensis*." *Journal of Apicultural Research* 2: 85–92.

Armstrong, D. (1969). "Dispositions Are Causes." *Analysis* 30: 23–26.

Austad, S. N. (1984). "A Classification of Alternative Reproductive Behaviors and Methods for Field-Testing ESS Models." *American Zoologist* 24: 309–319.

Ayala, F. (1970). "Teleological Explanations in Evolutionary Biology." *Philosophy of Science* 37: 1–15.

Beatty, J. (1987). "Natural Selection and the Null Hypothesis." In J. Dupré, ed., *The Latest on the Best: Essays on Evolution and Optimality*. Cambridge: MIT Press: 53–76.

Beatty, J. (1994). "Ernst Mayr and the Proximate/Ultimate Distinction." *Biology and Philosophy* 9: 333–356.

Beatty, J. (1995). "The Evolutionary Contingency Thesis." In G. Wolters and J. G. Lennox, eds., *Concepts, Theories, and Rationality in the Biological Sciences*. Pittsburgh: University of Pittsburgh Press: 45–81.

Beatty, J. (1997). "Why Do Biologists Argue like They Do?" *Philosophy of Science* 64 (4): S432–S443.

Bechtel, W. (1986). *Integrating Scientific Disciplines*. Dordrecht: Martinus Nijhoff.

Bechtel, W., and R. C. Richardson (1993). *Discovering Complexity: Decomposition and Localization as Strategies in Scientific Research*. Princeton: Princeton University Press.

Belnap, N. D., and J. B. Steel, Jr. (1976). *The Logic of Questions and Answers*. New Haven: Yale University Press.

Bernstein, H., G. S. Byers, and R. E. Michod (1981). "The Evolution of Sexual Reproduction: The Importance of DNA Repair, Complementation, and Variation." *American Naturalist* 117: 537–549.

Beshers, S. N., and J. H. Fewell (2001). "Models of Division of Labor in Social Insects." *Annual Review of Entomology* 46: 413–440.

Bickle, J. (1997). *Psychoneural Reductionism: The New Wave*. Cambridge: MIT Press.

References

Bigelow, J., and Pargetter, R. (1987). "Functions." *Journal of Philosophy* 84 (4): 181–196.

Black, M. (1979). "More about Metaphor." In A. Ortony, ed., *Metaphor and Thought*. Cambridge: Cambridge University Press: 19–43.

Bock, W. J. (1977). "Adaptation and the Comparative Method." In M. K. Hecht, B. M. Hecht, and P. C. Goody, eds., *Major Patterns in Vertebrate Evolution*. Chicago: University of Chicago Press: 57–81.

Bonabeau, E., G. Theraulaz, and J. L. Deneubourg (1996). "Quantitative Study of the Fixed Threshold Model for the Regulation of Division of Labour in Insect Societies." *Proceedings of the Royal Society of London Series B – Biological Sciences* 263: 1565–1569.

Bonabeau, E., G. Theraulaz, J. Deneubourg, S. Aron, and S. Camazine (1997). "Self-Organization in Social Insects." *Trends in Ecology and Evolution* 12 (5): 188–193.

Bonner, J. T. (1969). *The Scale of Nature*. New York: Pegasus.

Bonner, J. T. (1988). *The Evolution of Complexity*. Princeton: Princeton University Press.

Bonner, J. T. (1998). "The Origins of Multicellularity." *Integrative Biology* 1: 27–36.

Bono, J. J. (1990). "Science, Discourse, and Literature: The Role/Rule of Metaphor in Science." In S. Peterfreund, ed., *Literature and Science: Theory and Practice*. Boston: Northeastern University Press: 59–90.

Boomsma, J. J., E. J. Fjerdingstad, and J. Frydenberg (1999). "Multiple Paternity, Relatedness and Genetic Diversity in *Acromyrmex* Leaf-Cutter Ants." *Proceedings of the Royal Society of London Series* B (266): 249–254.

Boorse, C. (1976). "Wright on Functions." *Philosophical Review* 85: 70–86.

Bourke, A. F. G. (1988). "Worker Reproduction in the Higher Eusocial *Hymenoptera*." *Quarterly Review of Biology* 63: 291–311.

Bourke, A. F. G., and N. R. Franks (1995). *Social Evolution in Ants*. Princeton: Princeton University Press.

Boyd, R. (1979). "Metaphor and Theory Change: What Is 'Metaphor' a Metaphor For?" In A. Ortony, ed., *Metaphor and Thought*. Cambridge: Cambridge University Press: 356–408.

Boyd, R. N., and Peter J. Richerson (1985). *Culture and the Evolutionary Process*. Chicago: University of Chicago Press.

Boyd, R. N., and Peter. J. Richerson (1991). "Culture and Co-Operation." In R. A. Hinde and J. Groebel, eds., *Cooperation and Prosocial Behavior*. Cambridge: Cambridge University Press: 27–48.

Brandon, R. (1978). "Adaptation and Evolutionary Theory." *Studies in History and Philosophy of Science* 9: 181–206.

Brandon, R. (1980). "A Structural Description of Evolutionary Theory." In P. D. Asquith and Ronald Giere, eds., *PSA 1980*. East Lansing: Philosophy of Science Association. 2: 427–439.

Brandon, R. (1981). "Biological Teleology: Questions and Explanations." *Studies in the History and Philosophy of Science* 12: 91–105.

Brandon, R. (1982). "The Levels of Selection." In P. Asquith and T. Nickles, eds., *PSA 1982*. East Lansing: Philosophy of Science Association. 1: 315–323.

Brandon, R. (1985a). "Adaptation Explanations: Are Adaptations for the Good of Replicators or Interactors?" In D. Depew and B. Weber, eds., *Evolution at a Crossroads: The*

New Biology and the New Philosophy of Science. Cambridge: MIT Press/A Bradford Book: 81–96.

Brandon, R. (1985b). "Phenotypic Plasticity, Cultural Transmission and Human Sociobiology." In J. H. Fetzer, ed., *Sociobiology and Epistemology*. Dordrecht: Reidel Publishing: 57–74.

Brandon, R. (1997). "Does Biology Have Laws? The Experimental Evidence." *Philosophy of Science* 64 (4): S444–S457.

Brandon, R., and Beatty, J. (1984). "Discussion: The Propensity Interpretation of 'Fitness' – No Interpretation Is No Substitute." *Philosophy of Science* 51: 342–347.

Breed, M. D., G. E. Robinson, and R. E. Page, Jr. (1990). "Division of Labor during Honey Bee Colony Defense." *Behavioral Ecology and Sociobiology* 27: 395–401.

Brower, J. (1960). "Experimental Studies of Mimicry." *American Naturalist* 44: 271–283.

Brower, L. P. (1988). "Avian Predation on the Monarch Butterfly and Its Implications for Mimicry Theory." *American Naturalist* 131: S4–S6.

Brown, J. H., and G. West, eds. (2000). *Scaling in Biology*. Oxford: Oxford University Press.

Burian, R. M. (1981). "Human Sociobiology and Genetic Determinism." *Philosophical Forum* 13: 43–66.

Burian, R. M. (1983). "Adaptation." In M. Grene, ed., *Dimensions of Darwinism: Themes and Counterthemes in Twentieth-Century Evolutionary Theory*. Cambridge: Cambridge University Press: 287–314.

Burian, R. M. (1986). "On Integrating the Study of Evolution and Development: Comments on Kauffman and Wimsatt." In W. Bechtel, ed., *Integrating Scientific Disciplines*. Dordrecht: Martinus Nijhoff: 209–228.

Buss, L. (1987). *The Evolution of Individuality*. Princeton: Princeton University Press.

Calderone, N. W. (1998). "Proximate Mechanisms of Age Polyethism in the Honey Bee *Apis mellifera L.*" *Apidologie* 29: 127–158.

Calderone, N. W., and R. E. Page, Jr. (1988). "Genotypic Variability in Age Polyethism and Task Specialisation in the Honey Bee, *Apis mellifera* (*Hymenoptera: Apidae*)." *Behavioral Ecology and Sociobiology* 22: 17–25.

Calderone, N. W., and R. E. Page, Jr. (1991). "The Evolutionary Genetics of Division of Labor in Colonies of the Honey Bee (*Apis mellifera*)." *American Naturalist* 138: 69–92.

Calderone, N. W., and R. E. Page, Jr. (1992). "Effects of Interactions among Genotypically Diverse Nestmates on Task Specializations by Foraging Honey Bees (*Apis mellifera*)." *Behavioral Ecology and Sociobiology* 30: 219–226.

Calderone, N. W., and R. E. Page, Jr. (1996). "Temporal Polyethism and Behavioural Canalization in the Honey Bee, *Apis mellifera*." *Animal Behavior* 51: 631–643.

Calderone, N. W., G. E. Robinson, and R. E. Page, Jr. (1989). "Genetic Structure and Division of Labor in Honeybee Societies." *Experientia* 45: 765–767.

Camazine, S. (1993). "The Regulation of Pollen Foraging by Honey Bees: How Foragers Assess the Colony's Need for Pollen." *Behavioral Ecology and Sociobiology* 32: 265–272.

Camazine, S., and J. Sneyd (1991). "A Model of Collective Nectar Source Selection by Honey Bees: Self-Organization Through Simple Rules." *Journal of Theoretical Biology* 149: 547–571.

Carnap, R. (1950). "Empiricism, Semantics, and Ontology." *Revue Internationale de Philosophie* 4: 20–40.

Carnap, R. (1956). "Empiricism, Semantics, and Ontology." *Revue Internationale de Philosophie* 4 (1950): 20–40. Reprinted in the Supplement to *Meaning and Necessity: A Study in Semantics and Modal Logic*, enlarged edition. Chicago: University of Chicago Press: 205–221.

Carrier, M. (1995). "Evolutionary Change and Lawlikeness: Beatty on Biological Generalization." In G. Wolters and J. G. Lennox, eds., *Concepts, Theories, and Rationality in the Biological Sciences*. Pittsburgh: University of Pittsburgh Press: 82–97.

Cartwright, N. D. (1980). "The Truth Doesn't Explain Much." *American Philosophical Quarterly* 17: 159–163.

Cartwright, N. D. (1982). *How the Laws of Physics Lie*. Oxford: Oxford University Press.

Cartwright, N. D. (1986). "Two Kinds of Teleological Explanation." In A. Donagan, A. N. Perovich, Jr., and M. V. Wedin, eds., *Human Nature and Natural Knowledge*. Dordrecht: Reidel: 201–210.

Cartwright, N. D. (1989). *Nature's Capacities and Their Measurement*. Oxford: Oxford University Press.

Cartwright, N. D. (1994). "Fundamentalism vs. the Patchwork of Laws." *Proceedings of the Aristotelian Society* 94: 279–292.

Cartwright, N. D. (1999). *The Dappled World: A Study of the Boundaries of Science*. Cambridge: Cambridge University Press.

Causey, R. L. (1977). *Unity of Science*. Dordrecht: Reidel.

Churchland, P. S. (1986). *Neurophilosophy Toward a Unified Science of the Mind-Brain*. Cambridge, Mass., and London: MIT Press.

Clutton-Brock, T. H., and P. H. Harvey (1977). "Primate Ecology and Social Organization." *Journal of Zoology London* 183: 1–39.

Clutton-Brock, T. H., P. H. Harvey, and B. Rudder (1977). "Sexual Dimorphism, Socioeconomic Sex Ratio and Body Weight in Primates." *Nature* 269: 797–800.

Collins, H. M. (1985). *Changing Order: Replication and Induction in Scientific Practice*. Beverly Hills and London: Sage.

Collins, H. M., ed. (1982). *The Sociology of Scientific Knowledge: A Sourcebook*. Bath: Bath University Press.

Comte, A. (1830 (1975)). "Course of Positive Philosophy." In G. Lenzer, ed., *Auguste Comte and Positivism: The Essential Writings*. New York: Harper: 71–86.

Cox, P. A. (1989). *The Elements: Their Origin, Abundance, and Distribution*. Oxford: Oxford University Press.

Crozier, R. H., and R. E. Page, Jr. (1985). "On Being the Right Size: Male Contributions and Multiple Mating in Social *Hymenoptera*." *Behavioral Ecology and Sociobiology* 18: 105–115.

Culver, D. A. (1999). Ecological Modeling of Lake Erie Trophic Dynamics. http://www.ijc.org/boards/cglr/modsum/culver.html.

Cummins, R. (1975). "Functional Analysis." *Journal of Philosophy* 72: 741–765.

Darden, L. (1986). "Relations among Fields in the Evolutionary Synthesis." In W. Bechtel, ed., *Integrating Scientific Disciplines*. Dordrecht: Nijhoff: 113–123.

Darden, L. (1991). *Theory Change in Science: Strategies from Mendelian Genetics*. New York: Oxford University Press.

References

Darden, L., and J. Cain (1989). "Selection Type Theories." *Philosophy of Science* 56: 106–129.

Darden, L., and N. Maull (1977). "Interfield Theories." *Philosophy of Science* 44: 43–64.

Darwin, C. (1857). *On the Origin of Species by Means of Natural Selection, or the Preservation of Favoured Races in the Struggle for Life*. London: John Murray.

Darwin, C. (1859). *On the Origin of Species*, 6th ed. New York: Macmillan, 1962.

Darwin, C. (1964). *On the Origin of Species*. Harvard: Harvard University Press.

Darwin, C. (1965). *The Expression of the Emotions in Man and Animals*. Chicago: University of Chicago Press.

Darwin, C. (1981). *The Descent of Man, and Selection in Relation to Sex* (1871). Princeton: Princeton University Press.

Dawkins, R. (1976). *The Selfish Gene*. Oxford: Oxford University Press.

Dawkins, R. (1978). "Replicator Selection and the Extended Phenotype." *Zeitschrift für Tierpsychologie* 47: 61–76.

Dawkins, R. (1982a). *The Extended Phenotype: The Gene as the Unit of Selection*. Oxford: Oxford University Press.

Dawkins, R. (1982b). "Replicators and Vehicles." In King's College Sociobiology Group, ed., *Current Problems in Sociobiology*. Cambridge: Cambridge University Press: 45–64.

Deneubourg, J., S. Goss, J. Pasteels, D. Fresneau, and J. Lachaud (1987). "Self-Organization Mechanisms in Ant Societies (II): Learning in Foraging and Division of Labor." *Experientia Supplementum* 54 (Behavior in Social Insects): 177–196.

Done, T. J. 1992. "Constancy and Change in Some Great Barrier Reef Coral Communities: 1980–1990." *American Zoologist* 32: 655–662.

Dreller, C., M. K. Fondrk, and R. E. Page, Jr. (1995). "Genetic Variability Affects the Behavior of Foragers in a Feral Honeybee Colony." *Naturwissenschaften* 82: 243–245.

Dretske, F. (1977). "Laws of Nature." *Philosophy of Science* 44: 248–268.

Dupré, J. (1983). "The Disunity of Science." *Mind* 92: 321–346.

Dupré, J. (1993). *The Disorder of Things: Metaphysical Foundations of the Disunity of Science*. Cambridge: Harvard University Press.

Dupré, J. (1996). "Metaphysical Disorder and Scientific Disunity." In P. Galison and D. Stump, eds., *The Disunity of Science*. Stanford: Stanford University Press: 101–117.

Durham, W. H. (1979). "Towards a Coevolution Theory of Human Biology and Culture." In N. A. Chagnon and W. Irons, eds., *Evolutionary Biology and Human Social Behavior: An Anthropological Perspective*. North Scituate, Mass.: Duxbury Press: 39–59.

Durham, W. H. (1991). *Coeveolution: Genes, Culture and Human Diversity*. Stanford: Stanford University Press.

Earman, J. (1984). "Laws of Nature: The Empiricist's Challenge." In R. J. Bogdan, ed., *D. M. Armstrong*. Kluwer Academic Publishers: 191–224.

Earman, J., J. Roberts, and S. Smith (2002). "Ceteris Paribus Lost." *Erkenntnis* 57, no. 3.

Eckert, C. D., M. L. Winston, and R. C. Ydenberg (1995). "The Relationship between Population Size, Amount of Brood, and Individual Foraging Behaviour in the Honey Bee, *Apis mellifera*." *Oecologia* 97: 248–255.

223

Egerton, F. N. (1973). "Changing Concepts of the Balance of Nature." *Quarterly Review of Biology* 48: 322–350.

Endler, J. A. (1986). *Natural Selection in the Wild*. Princeton: Princeton University Press.

Evans, H. E. (1966). "The Behavior Patterns of Solitary Wasps." *Annual Review of Entomology* 11: 123–154.

Evans, M. A., and H. E. Evans (1970). *William Morton Wheeler, Biologist*. Cambridge: Harvard University Press.

Fergusson, L. A., and M. L. Winston (1985). "The Effect of Worker Loss on Temporal Caste Structure in Colonies of the Honeybee (*Apis mellifera L.*)." *Canadian Journal of Zoology* 63: 777–780.

Fewell, J. H., and R. E. Page, Jr. (1993). "Genotypic Variation in Foraging Responses to Environmental Stimuli by Honey Bees, *Apis mellifera*." *Experientia* 49: 1106–1112.

Fewell, J. H., and R. E. Page, Jr. (1999). "The Emergence of Division of Labour in Forced Associations of Normally Solitary Ant Queens." *Evolutionary Ecology Research* 1: 537–548.

Fewell, J. H., and M. L. Winston (1992). "Colony State and Regulation of Pollen Foraging in the Honey Bee, *Apis mellifera L.*" *Behavioral Ecology and Sociobiology* 30: 387–393.

Feyerabend, P. K. (1975). *Against Method*. London: New Left Books.

Feyerabend, P. K. (1981). *Philosophical Papers*. Cambridge: Cambridge University Press.

Feynman, R. (1985). *QED, the Strange Theory of Light and Matter*. Harmondsworth: Penguin.

Feynman, R. (1995). *The Character of Physical Law*. Cambridge: MIT Press.

Fisher, R. A. (1930). *The Genetical Theory of Natural Selection*. Oxford: Clarendon Press.

Fodor, J. (1974). "Special Sciences." *Synthese* 28: 97–115.

Friedman, M. (1974). "Explanation and Scientific Understanding." *Journal of Philosophy* 71: 5–19.

Frumhoff, P. C., and J. Baker (1988). "A Genetic Component to Division of Labor within Honey Bee Colonies." *Nature* 333: 358–361.

Galison, P. (1996). "Introduction: The Context of Disunity." In P. Galison and D. J. S. Galison, eds., *The Disunity of Science: Boundaries, Contexts, and Power*. Palo Alto: Stanford University Press.

Galison, P., and D. Stump, eds. (1996). *The Disunity of Science*. Stanford: Stanford University Press.

Garfinkel, A. (1981). *Forms of Explanation*. New Haven: Yale University Press.

Gelfand, A. E., and C. C. Walker (1984). *Ensemble Modeling*. New York: Marcel Dekker.

Ghiselin, M. (1974). *The Economy of Nature and the Evolution of Sex*. Berkeley: University of California Press.

Giray, T., and G. E. Robinson (1994). "Effects of Intracolony Variability in Behavioral Development on Plasticity of Division of Labor in Honey Bee Colonies." *Behavioral Ecology and Sociobiology* 35: 13–20.

Giray, T., and G. E. Robinson (1996). "Common Endocrine and Genetic Mechanisms of Behavioral Development in Male and Worker Honey Bees and the Evolution of Division of Labor." *Proceedings of the National Academy of Science USA* 93: 11718–11722.

References

Glymour, C. (1982). "Causal Inference and Causal Explanation." In R. McLaughlin, ed., *What? Where? When? Why?* Dordrecht: Reidel: 179–191.

Glymour, C. (2001). *The Mind's Arrows: Bayes Nets and Graphical Causal Models in Psychology.* Cambridge: MIT Press.

Godfrey-Smith, P. (1994). "A Modern History Theory of Function." *Nous* 28 (3): 344–362.

Goodwin, B. C. (1994). *How the Leopard Changed Its Spots: The Evolution of Complexity.* New York: C. Scribner's Sons.

Goodwin, B. C., and P. Saunders, eds. (1992). *Theoretical Biology: Epigenetic and Evolutionary Order from Complex Systems.* Baltimore: Johns Hopkins University Press.

Gordon, D. M. (1996). "The Organization of Work in Social Insect Colonies." *Nature* 380: 121–124.

Gould, S. J. (1977). "Caring Groups and Selfish Genes." *Natural History* 86: 20–24.

Gould, S. J. (1978). "Sociobiology: The Art of Storytelling." *New Scientist* 80: 530–533.

Gould, S. J. (1980). "Evolutionary Biology of Constraint." *Daedalus* 109 (2): 39–52.

Gould, S. J. (1981). "The Units of Evolution: A Metaphysical Essay." In R. Harre and U. L. Jensen, eds., *The Philosophy of Evolution.* Brighton: Harvester Press: 22–44.

Gould, S. J. (1983). "The Hardening of the Modern Synthesis." In M. Grene, ed., *Dimensions of Darwinism.* Cambridge: Cambridge University Press: 72–93.

Gould, S. J. (1987a). "Freudian Slip." *Natural History* 92: 14–21.

Gould, S. J. (1987b). "Stephen Jay Gould Replies." *Natural History* 96: 4–6.

Gould, S. J. (1989). *Wonderful Life: Burgess Shale and the Nature of History.* New York: W. W. Norton.

Gould, S. J., and E. S. Vrba (1982). "Exaptation – A Missing Term in the Science of Form." *Paleobiology* 8: 4–15.

Gould, S. J., and R. C. Lewontin (1979). "The Spandrels of San Marco and the Panglossian Paradigm: A Critique of the Adaptationist Program." *Proceedings of the Royal Society of London* 205: 581–598.

Grant, V. (1963). *The Origin of Adaptations.* New York: Columbia University Press.

Guilford, T. (1988). "The Evolution of Conspicuous Coloration." *American Naturalist* 131: S7–S21.

Guzmán-Novoa, E., and N. E. Gary (1993). "Genotypic Variability of Components of Foraging Behavior in Honey Bees (*Hymenoptera: Apidae*)." *Journal of Economic Entomology* 86: 715–721.

Guzmán-Novoa, E., and R. E. Page, Jr. (1994). "Genetic Dominance and Worker Interactions Affect Honeybee Colony Defense." *Behavioral Ecology* 5: 91–97.

Guzmán-Novoa, E., R. E. Page, Jr., and N. E. Gary (1994). "Behavioral and Life-History Components of Division of Labor in Honey Bees (*Apis mellifera L.*)." *Behavioral Ecology and Sociobiology* 34: 409–417.

Hacking, I. (1996). "The Disunities of the Sciences." In P. Galison and D. Stump, eds., *The Disunity of Science.* Stanford: Stanford University Press: 37–74.

Harding, S. G. (1986). *The Science Question in Feminism.* Ithaca: Cornell University Press.

Harvey, P. H., and M. D. Pagel (1991). *The Comparative Method in Evolutionary Biology.* Oxford: Oxford University Press.

Hatcher, B., R. E. Johannes, and A. Robertson (1989). "Review of Research Relevant to the Conservation of Shallow Tropical Marine Ecosystems." *Oceanography and Marine Biology Annual* 27: 337–414.

Hellmich, R. L., J. M. Kulincevic, and W. C. Rothenbuhler (1985). "Selection for High and Low Pollen-Hoarding Honey Bees." *Journal of Heredity* 76: 155–158.

Hempel, C. G. (1945). "Studies in the Logic of Confirmation." *Mind* 54: 97–121.

Hempel, C. G. (1951). "General Systems Theory: A New Approach to Unity of Science 2. General Systems Theory and the Unity of Science." *Human Biology* 23: 313.

Hempel, C. G. (1965a). *Aspects of Scientific Explanation.* New York: Free Press.

Hempel, C. G. (1965b). "The Logic of Functional Analysis." In *Aspects of Scientific Explanation.* New York: Free Press: 297–330.

Hempel, C. G. (1965c). "Studies in the Logic of Confirmation." In *Aspects of Scientific Explanation.* New York: Free Press.

Hesse, M. B. (1966). *Models and Analogies in Science.* Notre Dame, Ind.: University of Notre Dame Press.

Hesse, M. B. (1974). *The Structure of Scientific Inference.* London: Macmillan.

Hesse, M. B. (1986). "Theories, Family Resemblances and Analogy." In D. H. Helman, ed., *Analogical Reasoning.* Dordrecht: Kluwer: 317–340.

Holland, J. H. (1995). *Hidden Order: How Adaptation Builds Complexity.* Reading, Mass.: Addison-Wesley.

Hölldobler, B., and E. O. Wilson (1990). *The Ants.* Cambridge: Harvard University Press.

Holling, C. S. (1986). "The Resilience of Terrestrial Ecosystems: Local Surprise and Global Change." In W. M. Clark and R. E. Munn, eds., *Sustainable Development in the Biosphere.* Cambridge: Cambridge University Press: 292–320.

Horan, B. L. (1989). "Functional Explanations in Sociobiology." *Biology and Philosophy* 4: 131–158.

Horgan, J. (1996). *The End of Science.* Reading, Mass.: Addison-Wesley.

Huang, Z.-Y., and G. E. Robinson (1992). "Honeybee Colony Integration: Worker-Worker Interactions Mediate Hormonally Regulated Plasticity in Division of Labor." *Proceedings of the National Academy of Science USA* 89: 11726–11729.

Huang, Z.-Y., and G. E. Robinson (1996). "Regulation of Honey Bee Division of Labor by Colony Age Demography." *Behavioral Ecology and Sociobiology* 39: 147–158.

Hull, D. L. (1981). "The Units of Evolution: A Metaphysical Essay." In U. L. Jensen and R. Harre, eds., *The Philosophy of Evolution.* Brighton: Harvester Press: 22–44.

Hull, D. L. (1987). "On Human Nature." In A. Fine and Peter Machamer, eds., *PSA 1986.* East Lansing, Mich.: Philosophy of Science Association. 2: 3–13.

Hull, D. L. (1989). *The Metaphysics of Evolution.* Baltimore: Johns Hopkins University Press.

Hunt, G. J., R. E. Page, Jr., M. K. Fondrk, and C. J. Dullum (1995). "Major Quantitative Trait Loci Affecting Honey Bee Foraging Behavior." *Genetics* 141: 1537–1545.

Irons, W. (1979). "Cultural and Biological Success." In N. A. Chagon and W. Irons, eds., *Evolutionary Biology and Human Social Behavior: An Anthropological Perspective.* North Scituate, Mass.: Duxbury Press: 257–272.

James, W. (1950). *Principles of Psychology.* New York: Dover.

Jamieson, I. G. (1986). "The Functional Approach to Behavior: Is It Useful?" *American Naturalist* 127: 195–208.

References

Jamieson, I. G. (1989). "Levels of Analysis or Analyses at the Same Level." *Animal Behavior* 37: 696–697.

Kauffman, S. (1984). "Emergent Properties in Random Complex Automata." *Physica* 10D: 145–156.

Kauffman, S. (1987). "Developmental Logic and Its Evolution." *BioEssays* 6: 82–87.

Kauffman, S. (1993). *The Origins of Order*. Oxford: Oxford University Press.

Kauffman, S. (1995). *At Home in the Universe: The Search for Laws of Self-Organization and Complexity*. New York: Oxford University Press.

Keller, E. F., and L. A. Segel (1970). "Initiation of Slime-Mold Aggregation Viewed as an Instability." *Journal of Theoretical Biology* 26: 399–415.

Kessen, R. H. (2001). *Dictyostelium: Evolution, Cell Biology, and the Development of Multicellularity*. Cambridge: Cambridge University Press.

Kettlewell, H. B. D. (1956). "Further Selection Experiments on Industrial Melanism in the Lepidoptera." *Heredity* 10: 287–301.

Kim, J. (1990). "Supervenience as a Philosophical Concept." *Metaphilosophy* 21: 1–27.

Kitcher, P. (1981). "Explanatory Unification." *Philosophy of Science* 48: 507–531.

Kitcher, P. (1985). *Vaulting Ambition: Sociobiology and the Quest for Human Nature*. Cambridge and London: MIT Press.

Kitcher, P. (1990). "The Division of Cognitive Labor." *Journal of Philosophy* 87 (1): 5–22.

Koenig, W. D., and R. L. Mumme (1990). "Levels of Analysis and the Functional Significance of Helping Behaving." In M. Bekoff and D. Jamieson, eds., *Interpretation and Explanation in the Study of Behavior*, vol. 2: *Explanation, Evolution, and Adaptation*. Boulder: Westview Press: 268–303.

Koonce, J. F., and A. B. Locci. (1999). Resolution of Issues of Scope and Detail in the Development of the Lake Erie Ecological Model. http://www.ijc.org/boards/cglr/modsum/koonce.html.

Krebs, C. J. (1994). *Ecology: The Experimental Analysis of Distribution and Abundance*. New York: Harper Collins.

Kuhn, T. S. (1962). *The Structure of Scientific Revolutions*. Chicago: University of Chicago Press.

Kuhn, T. S. (1979). "Metaphor in Science." In A. Ortony, ed., *Metaphor and Thought*. Cambridge: Cambridge University Press: 409–419.

Lakatos, I. (1970). "The Methodology of Scientific Research Programmes." In I. Lakatos and A. Musgrave, eds., *Criticism and the Growth of Knowledge*. Cambridge: Cambridge University Press: 91–196.

Lakatos, I. (1971). "Falsificationism and the Methodology of Scientific Research Programmes." In I. Lakatos and Alan Musgrave, eds., *Criticism and the Growth of Knowledge*. Cambridge: Cambridge University Press: 91–195.

Lakatos, I. (1978). *The Methodology of Scientific Research Programmes*. Cambridge: Cambridge University Press,.

Lange, M. (2000). *Natural Laws in Scientific Practice*. Oxford: Oxford University Press.

Lange, M. (2002). "Who's Afraid of Ceteris Paribus Laws: Or: How I Learned to Stop Worrying and Love Them." *Erkenntnis* 57, no. 3.

Latour, B. (1988). "The Politics of Explanation." In S. Woolgar, ed., *Knowledge and Reflexivity: New Frontiers in the Sociology of Knowledge*. London: Sage: 155–176.

Lenoir, T. (1989). *The Strategy of Life*. Chicago: University of Chicago Press.

Levin, S. A. (1992). "The Problem of Pattern and Scale in Ecology: The Robert H. MacArthur Award Lecture." *Ecology* 73: 1943–1967.

Levins, R. (1968). *Evolution in Changing Environments*. Princeton: Princeton University Press.

Lewontin, R. C. (1970). "The Units of Selection." *Annual Review of Ecology and Systematics* 1: 1–18.

Lewontin, R. C. (1974). "The Analysis of Variance and the Analysis of Causes." *American Journal of Human Genetics* 26: 400–411.

Lewontin, R. C. (1978). "Adaptation." *Scientific American* 238: 157–169.

Lewontin, R. C., and L. C. Dunn (1960). "The Evolutionary Dynamics of a Polymorphism in the House Mouse." *Genetics* 45: 705–722.

Liebold, M. A., J. M. Chase, J. B. Shurin, and A. L. Downing (1997). "Species Turnover and the Regulation of Trophic Structure." *Annual Review of Ecology and Systematics* 28: 467–494.

Lindauer, M. (1961). *Communication among Social Bees*. Cambridge: Harvard University Press.

Longino, H. (1990). *Science as Social Knowledge*. Princeton: Princeton University Press.

Lumsden, C. (1982). "The Social Regulation of Physical Caste: The Superorganism Revived." *Journal of Theoretical Biology*: 749–781.

MacDonald, G. (1986). "The Possibility of the Dis-Unity of Science." In G. MacDonald, ed., *Fact, Science and Morality*. Oxford: Basil Blackwell: 219–246.

Mach, E. (1883). *The Science of Mechanics*, 6th ed. Chicago: Open Court Publishing, 1960.

Maeterlink, M. (1912). *The Life of the Bee*. New York: Dodd, Mead.

Maienschein, J. (1981). "Shifting Assumptions in American Biology." *Journal of the History of Biology* 14: 89–113.

Martin, J., and R. Harre (1982). "Metaphors in Science." In D. S. Miall, ed., *Metaphor: Problems and* Perspectives. Sussex: Harvester Press and Humanities Press: 89–106.

Maynard Smith, J. (1985). "The Birth of Sociobiology." *New Scientist* (26 September): 48–50.

Maynard Smith, J. (1987). "How to Model Evolution." In J. Dupré, *The Latest on the Best: Essays on Evolution and Optimality*. Cambridge: MIT Press: 119–131.

Maynard Smith, J. (1989). "Weismann and Modern Biology." *Oxford Surveys in Evolutionary Biology* 6: 1–12.

Maynard Smith, J. (1990). "Triumphs of Colonialism." *New York Review of Books* (27 September): 36–37.

Maynard Smith, J., R. Burian, S. Kauffman, R. Alberch, J. Campbell, B. Goodwin, R. Lande, D. Raup, and L. Wolpert (1985). "Developmental Constraints and Evolution." *Quarterly Review of Biology* 60: 265–286.

Mayr, E. (1961). "Cause and Effect in Biology." *Science* 134: 1501–1506.

Mayr, E. (1963). *Animal Species and Evolution*. Cambridge: Harvard University Press.

Mayr, E. (1974a). "Behavior Programs and Evolutionary Strategies." *American Scientist* 62: 650–659.

Mayr, E. (1974b). "Teleological and Teleonomic, a New Analysis." *Boston Studies in the Philosophy of Science* 14: 91–117.

Mayr, E. (1978). "Evolution." *Scientific American* 239: 46–55.

References

Mayr, E. (1982). *The Growth of Biological Thought: Diversity, Evolution, and Inheritance.* Cambridge: Belknap Press.

Michener, C. D. (1974). *The Social Bees.* Cambridge: Belknap Press of Harvard University Press.

Michener, C. D. (1975). *The Social Behavior of the Bees.* Cambridge: Belknap Press.

Miller, J. G. (1978). *Living Systems.* New York: McGraw-Hill.

Millikan, R. G. (1984). *Language, Thought, and Other Biological Categories.* Cambridge: MIT Press.

Millikan, R. G. (1989). "In Defense of Proper Functions." *Philosophy of Science* 56: 288–302.

Mills, S., and J. Beatty (1979). "The Propensity Interpretation of Fitness." *Philosophy of Science* 46: 263–286.

Mitchell, S. D. (1987a). "Can Evolution Adapt to Cultural Selection?" In A. Fine and P. Machamer, eds., *PSA 1986.* East Lansing, Mich.: Philosophy of Science Association. 2: 87–96.

Mitchell, S. D. (1987b). "Competing Units of Selection? A Case of Symbiosis." *Philosophy of Science* 54: 351–367.

Mitchell, S. D. (1987c). "'Why' Functions (in Evolutionary Biology and Cultural Anthropology)." In *History and Philosophy of Science.* Pittsburgh: University of Pittsburgh: 201.

Mitchell, S. D. (1989). "The Causal Background for Functional Explanations." *International Studies in the Philosophy of Science* 3 (2): 213–230.

Mitchell, S. D. (1990). "Units of Behavior in Evolutionary Explanations." In D. Jamieson and M. Bekoff, eds., *Interpretation and Explanation in the Study of Animal Behavior.* Boulder: Westview Press: 63–83.

Mitchell, S. D. (1992). "On Pluralism and Competition in Evolutionary Explanations." *American Zoologist* 32: 135–144.

Mitchell, S. D. (1993). "Dispositions or Etiologies? A Comment on Bigelow and Pargetter." *Journal of Philosophy* 90: 249–259.

Mitchell, S. D. (1995a). "Function, Fitness and Disposition." *Biology and Philosophy* 10: 39–54.

Mitchell, S. D. (1995b). "The Superorganism Metaphor: Then and Now." In E. M. S. Maasen and P. Weingart, eds., *Biology as Society, Society as Biology: Metaphors.* Amsterdam: Kluwer Academic Publishers: 231–248.

Mitchell, S. D. (1997). "Pragmatic Laws." *Philosophy of Science* 64: S468– S479.

Mitchell, S. D. (2000). "Dimensions of Scientific Law." *Philosophy of Science* 67: 242– 265.

Mitchell, S. D. (2002a). "Ceteris Paribus – An Inadequate Representation for Biological Contingency." *Erkenntnis* 57, no. 3: 329–350.

Mitchell, S. D. (2002b). "Contingent Generalizations: Lessons from Biology." In R. Mayntz, ed., *Akteure, Mechanismen, Modelle: Zur Theoriefähigkeit makrosozialer Analysen.* Frankfurt: Campus: 179–195.

Mitchell, S. D. (2002c). "Integrative Pluralism." *Biology and Philosophy* 17: 55–70.

Mitchell, S. D., and Robert E. Page, Jr. (1992). *Idiosyncratic Paradigms and the Revival of the Superorganism.* Bielefeld: Center for Interdisciplinary Studies.

Mitchell, S. D., L. Daston, G. Gigerenzer, N. Sesardic, and P. Sloep (1997). "The How's and Why's of Interdisciplinarity." In P. Weingart, S. D. Mitchell, P. Richerson, and S. Maasen, eds., *Human by Nature: Between Biology and the Social Sciences.* Mahwah, N.J.: Erlbaum Press: 103–150.

Mitman, G. (1995). "Defining the Organism in the Welfare State: The Politics of Individuality in American Culture, 1890–1950." In E. M. S. Maasen and P. Weingart, eds., *Sociology of the Sciences Yearbook 18 (1994): Biology as Society, Society as Biology: Metaphors.* Amsterdam: Kluwer Academic Publishers: 249–280.

Morgan, C. L. (1895). *An Introduction to Comparative Psychology.* London: Walter Scott.

Morrison, M. (2000). *Unifying Scientific Theories: Physical Concepts and Mathematical Structures.* Cambridge: Cambridge University Press.

Moser, P. K., and J. D. Trout, eds. (1995). *Contemporary Materialism.* London and New York: Routledge.

Murray, J. D. (1988). "How the Leopard Gets Its Spots." *Scientific American* 259: 80–87.

Nagel, E. (1977). "Teleology Revisited." *Journal of Philosophy* 74: 261–301.

National Science Foundation (1995). *NSF in a Changing World: The National Science Foundation's Strategic Plan.* http://www.nsf.gov/nsf/nsfpubs/straplan/contents.htm.

Neander, K. (1991). "Functions as Selected Effects: The Conceptual Analyst's Defense." *Philosophy of Science* 58: 168–184.

Neurath, O. (1987). "Unified Science and Physicalism." In B. McGuiness, ed., *Unified Science* (1932). Dordrecht: Reidel.

Nicolis, G., and I. Prigogine (1989). *Exploring Complexity: An Introduction.* New York: W. H. Freeman.

Nicolis, S. C., and J. L. Deneubourg (1999). "Emerging Patterns and Food Recruitment in Ants: An Analytical Study." *Journal of Theoretical Biology* 198: 575–592.

Noonan, K. C. (1986). "Recognition of Queen Larvae by Worker Honey Bees (*Apis mellifera*)." *Ethology* 73: 295–306.

Nordmann, A. (1990). "Persistent Propensities: Portrait of a Familiar Controversy." *Biology and Philosophy* 5: 379–399.

Oldroyd, B. P., T. E. Rinderer, and S. M. Buco. (1990). "Nepotism in Bees." *Nature* 346: 706–708.

Oldroyd, B. P., T. E. Rinderer, and S. M. Buco (1991). "Honey Bees Dance with Their Super Sisters." *Animal Behavior* 42: 121–129.

Oppenheim, P., and Hilary Putnam (1958). "Unity of Science as a Working Hypothesis." In M. S. Herbert Feigl and Grover Maxwell, eds., *Concepts, Theories and the Mind-Body Problem.* Minnesota Studies in the Philosophy of Science. Minneapolis: University of Minnesota Press. 2: 3–36.

Oreskes, N., K. Shrader-Frechette, and K. Belitz (1994). "Verification, Validation, and Confirmation of Numerical Models in the Earth Sciences." *Science* 263 (5147): 641–646.

Ortony, A. (1979). *Metaphor and Thought.* Cambridge: Cambridge University Press.

Oster, G., and E. O. Wilson (1978). *Caste and Ecology in the Social Insects.* Princeton: Princeton University Press.

Oster, G., and J. Murray (1989). "Pattern Formation Models and Developmental Constraints." *Journal of Experimental Zoology* 251: 186–202.

References

Page, R. E., Jr. (1986). "Sperm Utilization in Social Insects." *Annual Review of Entomology* 31: 297–320.

Page, R. E., Jr. (1997). "The Evolution of Insect Societies." *Endeavour* 21: 114–120.

Page, R. E., Jr., and J. Erber (2002). "Levels of Behavioral Organization and the Evolution of Division of Labor." *Naturwissenschaften* 89: 91–106.

Page, R. E., Jr., and E. H. Erickson (1984). "Selective Rearing of Queens by Worker Honey Bees: Kin or Nestmate Recognition?" *Annals of the Entomological Society of America* 77: 578–580.

Page, R. E., Jr., and E. H. Erickson (1988). "Reproduction by Worker Honey Bees *Apis mellifera L.*" *Behavioral Ecology and Sociobiology* 23 (2): 117–126.

Page, R. E., Jr., and M. K. Fondrk (1995). "The Effects of Colony-Level Selection on the Social Organization of Honey Bee (*Apis mellifera L.*) Colonies: Colony-Level Components of Pollen Hoarding." Behavioral Ecology and Sociobiology 36: 135–144.

Page, R. E., Jr., and R. A. Metcalf (1982). "Multiple Mating, Sperm Utilization, and Social Evolution." *American Naturalist* 119: 263–281.

Page, R. E., Jr., and S. D. Mitchell (1991). "Self Organization and Adaptation in Insect Societies." In A. Fine, M. Forbes, and L. Wessels, eds., *PSA 1989*. East Lansing: Philosophy of Science Association. 2: 289–298.

Page, R. E., Jr., and S. D. Mitchell (1998). "Self Organization and the Evolution of Division of Labor." *Apidologie* 29: 101–120.

Page, R. E., Jr., and G. E. Robinson (1990). "Nepotism in Bees." *Nature* 346: 708.

Page, R. E., Jr., and G. E. Robinson (1991). "The Genetics of Division of Labour in Honey Bee Colonies." *Advances in Insect Physiology* 23: 117–169.

Page, R. E., Jr., G. E. Robinson, and M. K. Fondrk (1989a). "Genetic Specialists, Kin Recognition, and Nepotism in Honey-Bee Colonies." *Nature* 338: 576–579.

Page, R. E., Jr., G. E. Robinson, N. W. Calderone, and W. C. Rothenbuhler (1989b). "Genetic Structure, Division of Labor, and the Evolution of Insect Societies." In M. D. Breed and R. E. Page, eds., *The Genetics of Social Evolution*. Boulder: Westview Press: 15–29.

Page, R. E., Jr., G. E. Robinson, D. S. Britton, and M. K. Fondrk (1992). "Genotypic Variability for Rates of Behavioral Development in Worker Honeybees (*Apis mellifera*)." *Behavioral Ecology* 4: 173–180.

Page, R. E., Jr., G. E. Robinson, M. K. Fondrk, and M. E. Nasr (1995a). "Effects of Worker Genotypic Diversity on Honey Bee Colony Development and Behavior (*Apis mellifera L.*)." *Behavioral Ecology and Sociobiology* 26: 387–396.

Page, R. E., Jr., K. D. Waddington, G. J. Hunt, and M. K. Fondrk (1995b). "Genetic Determinants of Honey Bee Foraging Behavior." *Animal Behavior* 50: 1617–1625.

Page, R. E., Jr., J. Erber, and M. K. Fondryk (1998). "The Effect of Genotype on Response Thresholds to Sucrose and Foraging Behavior of Honey Bees (*Apis mellifera L.*)." *Journal of Comparative Physiology A* 182: 489–500.

Pankiw, T., K. D. Waddington, and R. E. Page (2001). "Modulation of Sucrose Response Thresholds in Honey Bees (*Apis mellifera L.*): Influence of Genotype, Feeding, and Foraging Experience." *Journal of Comparative Physiology A* 187: 293–301.

Pereira, H. M., and D. M. Gordon (2001). "A Trade-Off in Task Allocation Between Sensitivity to the Environment and Response Time." *Journal of Theoretical Biology* 208: 165–184.

Pietroski, P., and G. Rey (1995). "When Other Things Aren't Equal: Savings 'Ceteris Paribus' Laws from Vacuity." *British Journal for the Philosophy of Science* 46 (1): 81–110.

Pool, R. (1997). "Why Nature Loves Economies of Scale." *New Scientist* 154 (2077): 16.

Prior, E. (1985). *Dispositions*. Atlantic Highlands, N.J.: Humanities Press.

Prior, E., R. Pargetter, and F. Jackson (1982). "Three Theses About Dispositions." *American Philosophical Quarterly* 19: 251–257.

Raff, R. A. (1996). *The Shape of Life*. Chicago: University of Chicago Press.

Raff, R. A., G. A. Wray, and J. J. Henry (1991). "Implications of Radical Evolutionary Changes in Early Development for Concepts of Developmental Constraints." In L. Warren and M. Meselson, eds., *New Perspectives on Evolution*. New York: Wiley-Liss: 189–207.

Rappaport, R. (1968). *Pigs for the Ancestors*. New Haven: Yale University Press.

Ratnieks, F. L. W., and P. K. Visscher (1989). "Worker Policing in the Honeybee." *Nature* 342: 796–797.

Reeve, H. K., and P. W. Sherman (1993). "Adaptation and the Goals of Evolutionary Research." *Quarterly Review of Biology* 68 (1): 1–32.

Reif, F. (1987). *Statistical Physics*. New York: McGraw-Hill.

Resnik, D. (1988). "Survival of the Fittest: Law of Evolution or Law of Probability?" *Biology and Philosophy* 3: 349–362.

Richerson, P. J., and R. N. Boyd (1987). "Simple Models of Complex Phenomena: The Case of Cultural Evolution." In J. Dupré, ed., *The Latest on the Best: Essays on Evolution and Optimality*. Cambridge: MIT Press: 27–54.

Ricklefs, R. (1973). *The Economy of Nature*. New York: Chiron Press.

Robinson, G. E. (1992). "Regulation of Division of Labor in Insect Societies." *Annual Review of Entomology* 37: 637–665.

Robinson, G. E., and R. E. Page, Jr. (1988). "Genetic Determination of Guarding and Undertaking in Honey-Bee Colonies." *Nature* 333: 356–358.

Robinson, G. E., and R. E. Page, Jr. (1989a). "Genetic Basis for Division of Labor in an Insect Society." In M. D. Breed and R. E. Page, eds., *The Genetics of Social Evolution*. Boulder: Westview Press: 61–80.

Robinson, G. E., and R. E. Page, Jr. (1989b). "Genetic Determination of Nectar Foraging, Pollen Foraging, and Nest-Site Scouting in Honey Bee Colonies." *Behavioral Ecology and Sociobiology* 24: 317–323.

Robinson, G. E., and R. E. Page, Jr. (1995). "Genotypic Constraints on Plasticity for Corpse Removal in Honey Bee Colonies." *Animal Behavior* 49: 867–876.

Robinson, G. E., R. E. Page, Jr., C. Strambi, and A. Strambi (1989). "Hormonal and Genetic Control of Behavioral Integration in Honey Bee Colonies." *Science* 246: 109–112.

Robinson, G. E., R. E. Page, Jr., and M. K. Fondrk (1990). "Intracolonial Behavioral Variation in Worker Oviposition, Oophagy, and Larval Care in Queenless Honey Bee Colonies." *Behavioral Ecology and Sociobiology* 26: 315–323.

Robinson, G. E., R. E. Page, Jr., and N. Arensen (1994a). "Genotypic Differences in Brood Rearing in Honey Bee Colonies: Context Specific?" *Behavioral Ecology and Sociobiology* 34: 125–137.

Robinson, G. E., R. E. Page, Jr., and Z.-Y. Huang (1994b). "Temporal Polyethism in Social Insects Is a Developmental Process." *Animal Behavior* 48: 467–469.

Rosenberg, A. (1982). "Discussion: On the Propensity Interpretation of Fitness." *Philosophy of Science* 49: 268–273.

Rosenberg, A. (1985). *The Structure of Biological Science*. Cambridge: Cambridge University Press.

Ross, K. G. (1988). "Reproductive and Social Structure in Polygamous Fire Ant Colonies." In M. D. Breed and R. E. Page, eds., *The Genetics of Social Evolution*. Boulder: Westview Press: 149–162.

Rothenbuhler, W. C., and R. E. Page, Jr. (1989). "Genetic Variability for Temporal Polyethism in Colonies Consisting of Similarly-Aged Worker Honey Bees." *Apidologie* 29: 433–437.

Ruse, M. (1988). "Molecules to Men: Evolutionary Biology and Thoughts of Progress." In M. H. Nitecki, ed., *Evolutionary Progress*. Chicago: University of Chicago Press.

Salmon, W. C. (1971). *Statistical Explanation and Statistical Relevance*. Pittsburgh: University of Pittsburgh Press.

Salmon, W. C. (1990). *Four Decades of Scientific Explanation*. Minneapolis: University of Minnesota Press.

Salthe, S. N. (1985). *Evolving Hierarchical Systems: Their Structure and Representation*. New York: Columbia University Press.

Salthe, S. N. (1993). *Development and Evolution: Complexity and Change in Biology*. Cambridge: MIT Press.

Scheiner, R., R. E. Page, Jr., and J. Erber (2001). "Responsiveness to Sucrose Affects Tactile and Olfactory Learning in Preforaging Honey Bees of Two Genetic Strains." *Behavioral and Brain Research* 120: 67–73.

Schouten, M. K. D., and H. L. De Jong (1999). "Reduction, Elimination, and Levels: The Case of the LTP-Learning Link." *Philosophical Psychology* 12: 237–262.

Seeley, T. (1982). "Adaptive Significance of the Age Polyethism Schedule in Honey Bee Colonies." *Behavior, Ecology and Sociobiology* 11: 287–293.

Seeley, T. (1985). *Honey Bee Ecology: A Study in Adaptation in Social Life*. Princeton: Princeton University Press.

Seeley, T. (1989). "The Honey Bee Colony as a Superorganism." *American Scientist* 77: 546–553.

Seeley, T. (1995). *Wisdom of the Hive*. Cambridge: Harvard University Press.

Seeley, T., and C. A. Tovey (1994). "Why Search Time to Find a Food-Storer Bee Accurately Indicates the Relative Rates of Nectar Collecting and Nectar Processing in Honey Bee Colonies." *Animal Behavior* 47 (2): 311–316.

Selander, R. K. (1972). "Sexual Selection and Dimorphism in Birds." In B. G. Campbell, ed., *Sexual Selection and the Descent of Man 1871–1971*. Chicago: Aldine: 180–230.

Shakespeare, W. (1624). *Hamlet*. In W. J. Craig, ed., *The Oxford Shakespeare: The Complete Works of William Shakespeare*. London: Oxford University Press, 1914.

Sherman, P. W. (1988). "The Levels of Analysis." *Animal Behavior* 36: 616–619.

Sherman, P. W. (1989). "The Clitoris Debates and the Levels of Analysis." *Animal Behavior* 37: 697–698.

Shields, W. M., and L. M. Shields (1983). "Forcible Rape: An Evolutionary Perspective." *Ethology and Sociobiology* 4: 115–136.

Simon, H. (1962). "The Architecture of Complexity." In *The Sciences of the Artificial*. Cambridge: MIT Press: 192–229.

Simon, H. (1970). *Sciences of the Artificial*. Cambridge: MIT Press.

Simon, H. (1981). *The Science of the Artificial*. Cambridge: MIT Press.

Skipper, R. (1999). "Selection and the Extent of Explanatory Unification." *Philosophy of Science: Supplement* 66: S196–209.

Skyrms, B. (1980). *Causal Necessity: A Pragmatic Investigation of the Necessity of Laws*. New Haven: Yale University Press.

Sloep, P. (1993). "Methodology Revitalized?" *British Journal for the Philosophy of Science* 44 (2): 231–251.

Smart, J. J. C. (1968). *Between Science and Philosophy: An Introduction to the Philosophy of Science*. New York: Random House.

Sober, E. (1984a). *Conceptual Issues in Evolutionary Biology*. Cambridge: MIT Press.

Sober, E. (1984b). "Force and Disposition in Evolutionary Theory." In C. Hookway, ed., *Minds, Machines and Evolution*. Cambridge: Cambridge University Press: 43–61.

Sober, E. (1984c). *The Nature of Selection*. Cambridge and London: MIT Press.

Sober, E. (1987). "What Is Adaptationism?" In J. Dupré, ed., *The Latest on the Best: Essays on Evolution and Optimality*. Cambridge: MIT Press: 105–118.

Sober, E. (1997). "Two Outbreaks of Lawlessness in Recent Philosophy of Biology." *Philosophy of Science* 64 (4): S458–S467.

Soltis, J., R. N. Boyd, and P. J. Richerson (1991). *Can Group-Functional Behaviors Evolve by Cultural Group Selection? An Empirical Test*. Bielefeld: Center for Interdiscipinary Research.

Spirtes, P., C. Glymour, and R. Scheines (1993). *Causation, Prediction, and Search*. New York: Springer-Verlag.

Stearns, S. C. (1992). *The Evolution of Life Histories*. Oxford: Oxford University Press.

Stepan, N. (1986). "Race and Gender: The Role of Analogy in Science." *Isis* 77: 261–277.

Suppes, P. (1974). *Probabilistic Metaphysics*. Oxford: Blackwell.

Thornhill, R. (1979). "Review of Insect Behavior by R. W. Matthews and J. R. Matthews." *Quarterly Review of Biology* 54: 365–366.

Thornhill, R. (1980). "Rape in *Panorpa* Scorpionflies and a General Rape Hypothesis." *Animal Behavior* 28: 52–59.

Thornhill, R. (1984). "Scientific Methodology in Entomology." *Florida Entomologist* 67 (1): 74–96.

Thornhill, R., and J. Alcock (1983). *The Evolution of Insect Mating Systems*. Cambridge: Harvard University Press.

Thornhill, R., and N. W. Thornhill (1983). "Human Rape: An Evolutionary Analysis." *Ethology and Sociobiology* 4: 137–173.

Thornhill, R., and N. W. Thornhill (1987). "Human Rape: The Strengths of the Evolutionary Perspective." In C. Crawford, M. Smith, and D. Krebs, eds., *Sociobiology and Psychology: Ideas, Issues and Applications*. Hillsdale, N.J.: Lawrence Erlbaum Associates: 269–292.

Thornhill, R., and N. W. Thornhill (1989). "The Evolution of Psychological Pain." In R. Bell, ed., *Sociobiology and the Social Sciences*. Lubbock: Texas Tech University Press: 73–103.

Tinbergen, N. (1963). "On the Aims and Methods of Ethology." *Zeitschrift für Tierpsychologie* (*Journal of Comparative Ethology*) 20: 410–433.

Todes, D. (1989). *Darwin Without Malthus*. Oxford: Oxford University Press.

Tofts, C., and N. R. Franks (1992). "Doing the Right Thing: Ants, Honeybees and Naked Mole-Rats." *Trends in Ecology and Evolution* 7: 346–349.

Trivers, R. L. (1985). *Social Evolution*. Menlo Park, Calif.: Benjamin/Cummings.

Trump, R. F., V. C. Thompson, and W. C. Rothenbuhler (1967). "Behaviour Genetics of Nest Cleaning in Honeybees V. Effect of Previous Experience and Composition of Mixed Colonies on Response to Disease-Killed Brood." *Journal of Apicultural Research* 6: 127–131.

Van Frassen, B. C. (1980). *The Scientific Image*. Oxford: Clarendon Press.

Van Noppen, J. P., S. de Knop, and R. Jongen (1985). *Metaphors: A Bibliography of Post-1970 Publications*. Amsterdam: Benjamin.

Vayda, A. (1974). "Warfare in Ecological Perspective." *Annual Review of Ecology and Systematics* 5: 183–193.

Vayda, A. (1989). "Explaining Why Marings Fought." *Journal of Anthropological Research* 45 (2): 159–177.

Visscher, P. K. (1986). "Kinship Discrimination in Queen Rearing by Honey Bees (*Apis mellifera*)." *Behavioral Ecology and Sociobiology* 18: 453–460.

Visscher, P. K. (1989). "A Quantitative Study of Worker Reproduction in Honey Bee Colonies." *Behavioral Ecology and Sociobiology* 25: 247–254.

Vrba, E. (1989). "Levels of Selection and Sorting, with Special Reference to the Species Level." *Oxford Surveys of Evolutionary Biology* 6: 111–168.

Wade, M. J. (1978). "A Critical Review of the Models of Group Selection." *Quarterly Review of Biology* 53: 101–114.

Wade, M. J. (1980). "Kin Selection: Its Components." *Science* 210: 665–667.

Waters, C. K. (1986). "Natural Selection Without Survival of the Fittest." *Biology and Philosophy* 1: 207–225.

Waters, C. K. (1998). "Causal Regularities in the Biological World of Contingent Distributions." *Biology and Philosophy* 13: 5–36.

Weinert, R. (1995). *Laws of Nature*. Berlin: Walter de Gruyter.

West, G. B., J. H. Brown, and B. J. Enquist (1997). "A General Model for the Origin of Allometric Scaling Laws in Biology." *Science* 276: 122–130.

West-Eberhard, M. J. (1981). "Intragroup Selection and the Evolution of Insect Societies." In R. D. Alexander and D. W. Tinkle, eds., *Natural Selection and Social Behavior*. New York: Chiron Press: 3–17.

West-Eberhard, M. J.(1987). "Flexible Strategy and Social Evolution." In J. L. Brown and J. Kikkawa, eds., *Animal Societies: Theories and Facts*. Tokyo: Japan Science Society Press: 35–51.

Wheeler, W. M. (1911). "The Ant-Colony as an Organism." *Morphology* 22: 307–325.

Wheeler, W. M. (1923). *Social Life among the Insects*. New York: Harcourt, Brace.

Wheeler, W. M. (1928). *The Social Insects: Their Origin and Evolution*. New York: Harcourt, Brace.

Whitley, R. (1984). *Intellectual and Social Organisation of the Sciences*. Oxford: Clarendon Press.

Whitman, C. O. (1895). "Prefatory Note." *Biological Lectures 1894*: iii–vii.

Wickler, W. (1968). *Mimicry in Plants and Animals*. New York: McGraw-Hill.

References

Williams, G. C. (1966). *Adaptation and Natural Selection*. Princeton: Princeton University Press.

Wilson, D. S., and E. Sober (1989). "Reviving the Superorganism." *Journal of Theoretical Biology* 136: 337–356.

Wilson, E. O. (1968). "The Superorganism Concept and Beyond." In M. N. M. Chauvin and P. Grasse, eds., *L'effet de groupe chez les animaux*. Paris: Colloque internationaux du centre national de la recherche scientifique. No. 173: 27–39.

Wilson, E. O. (1971). *The Insect Societies*. Cambridge: Harvard University Press.

Wilson, E. O. (1978). *On Human Nature*. Cambridge: Harvard University Press.

Wilson, E. O. (1985a). The Principles of Caste Evolution." In B. Hölldobler and M. Lindauer, eds., *Experimental Behavioral Ecology and Sociobiology*. Sunderland, Mass.: Sinauer Associates: 307–324.

Wilson, E. O. (1985b). "The Sociogenesis of Insect Colonies." *Science* 228: 1489–1495.

Wimsatt, W. C. (1972). "Teleology and the Logical Structure of Function Statements." *Studies in History and Philosophy of Science* 3: 1–80.

Wimsatt, W. C. (1980). "Reductionistic Research Strategies and Their Biases in the Units of Selection Controversy." In T. Nickles, ed., *Scientific Discovery*. Dordrecht: Reidel. 2: 213–259.

Wimsatt, W. C. (1986). "Forms of Aggregativity." In P. Donagon and Wedin, eds., *Human Nature and Natural Knowledge*. Boston: D. Reidel: 259–291.

Wimsatt, W. C. (1987). "False Models as Means to Truer Theories." In M. H. Nitecki and A. Hoffman, eds., *Neutral Models in Biology*. Oxford: Oxford University Press: 23–55.

Wimsatt, W. C. (2000). "Emergence as Non-Aggregativity and the Biases of Reductionisms." *Foundations of Science* 5: 269–297.

Winston, M. L. (1987). *The Biology of the Honey Bee*. Cambridge: Harvard University Press.

Winston, M. L., and S. J. Katz (1982). "Foraging Differences between Cross-Fostered Honeybee Workers (*Apis mellifera*) of European and Africanized Races." *Behavioral Ecology and Sociobiology* 10: 125–129.

Winston, M. L., and K. N. Slessor (1998). "Honey Bee Primer Pheromones and Colony Organization: Gaps in Our Knowledge." *Apidologie* 29: 81–95.

Woodward, J. (1997). "Explanation, Invariance, and Intervention." *Philosophy of Science* 64: S26–S41.

Woodward, J. (2000). "Explanation and Invariance in the Special Sciences." *British Journal for the Philosophy of Science* 51: 197–255.

Woodward, J. (2001). "Law and Explanation in Biology: Invariance Is the Kind of Stability That Matters." *Philosophy of Science* 68: 1–20.

Woodward, J. (2002a). *A Theory of Explanation: Causation, Invariance, and Intervention*. Oxford: Oxford University Press.

Woodward, J. (2002b). "There Is No Such Thing as a Ceteris Paribus Law." *Erkenntnis* 57, no. 3.

Woolgar, S., ed. (1988). *Knowledge and Reflexivity: New Frontiers in the Sociology of Scientific Knowledge*. London: Sage.

Wouters, A. (1995). "Viability Explanation." *Biology and Philosophy* 10: 435–457.

References

Wright, L. (1973). "Functions." *Philosophical Review* 82 (2): 139–168.

Wright, L. (1976). *Teleological Explanations*. Los Angeles: University of California Press.

Wynne-Edwards, V. C. (1962). *Animal Dispersion in Relation to Social Behaviour*. Edinburgh: Oliver and Boyd.

Zimmering, S., L. Sandler, and B. Nicoletti (1970). "Mechanisms of Meiotic Drive." *Annual Review of Genetics* 4: 409–436.

Index